T0181163

# Clinical Data Analysis on a Pocket Calculator

Ton J. Cleophas • Aeilko H. Zwinderman

# Clinical Data Analysis on a Pocket Calculator

## Understanding the Scientific Methods of Statistical Reasoning and Hypothesis Testing

Second Edition

Ton J. Cleophas
Department of Medicine
Albert Schweitzer Hospital
Dordrecht, The Netherlands

Aeilko H. Zwinderman
Department Epidemiology and Biostatistics
Academic Medical Center
Amsterdam, The Netherlands

ISBN 978-3-319-80074-5 ISBN 978-3-319-27104-0 (eBook)
DOI 10.1007/978-3-319-27104-0

Springer Cham Heidelberg New York Dordrecht London

Softcover reprint of the hardcover 2nd edition 2016


Printed on acid-free paper

Springer International Publishing AG Switzerland is part of Springer Science+Business Media (www.springer.com)

# Prefaces to First Edition

## Book One

The time that statistical analyses, including analysis of variance and regression analyses, were analyzed by statistical analysts has gone for good, thanks to the availability of user-friendly statistical software. The teaching department, the educations committee, and the scientific committee of the Albert Schweitzer Hospital, Dordrecht, the Netherlands, are pleased to announce that since November 2009, the entire staff and personnel are able to perform statistical analyses with the help of SPSS Statistical Software in their offices through the institution's intranet.

It is our experience as master's and doctorate class teachers of the European College of Pharmaceutical Medicine (EC Socrates Project) that students are eager to master adequate command of statistical software for carrying out their own statistical analyses. However, students often lack adequate knowledge of basic principles, and this carries the risk of fallacies. Computers cannot think and can only execute commands as given. As an example, regression analysis usually applies independent and dependent variables, often interpreted as causal factors and outcome factors. For example, gender and age may determine the type of operation or the type of surgeon. The type of surgeon does not determine the age and gender. Yet, software programs have no difficulty to use nonsense determinants, and the investigator in charge of the analysis has to decide what is caused by what, because a computer cannot do a thing like that, although it is essential to the analysis.

It is our experience that a pocket calculator is very helpful for the purpose of studying the basic principles. Also, a number of statistical methods can be performed more easily on a pocket calculator, than using a software program.

Advantages of the pocket calculator method include the following:

1. You better understand what you are doing. The statistical software program is kind of a black box program.
2. The pocket calculator works faster, because far less steps have to be taken.

3. The pocket calculator works faster, because averages can be used.
4. With statistical software all individual data have to be included separately, a time-consuming activity in case of large data files.

Also, some analytical methods, for example, power calculations and required sample size calculations, are difficult on a statistical software program and easy on a pocket calculator. This book reviews the pocket calculator methods together with practical examples, both hypothesized and real patient data. The book was produced together with the similarly sized book *SPSS for Starters* from the same authors (edited by Springer, Dordrecht 2010). The two books complement one another. However, they can be studied separately as well.

Lyon, France Ton J. Cleophas
December 2010 Aeilko H. Zwinderman

## Book Two

The small book *Statistical Analysis of Clinical Data on a Pocket Calculator* edited in 2011 presented 20 chapters of cookbook-like step-by-step analyses of clinical data and was written for clinical investigators and medical students as a basic approach to the understanding and carrying out of medical statistics. It addressed subjects like the following:

1. Statistical tests for continuous/binary data
2. Power and sample size assessments
3. Calculation of confidence intervals
4. Calculating variabilities
5. Adjustments for multiple testing
6. Reliability assessments of qualitative and quantitative diagnostic tests

This book is a logical continuation and reviews additional pocket calculator methods that are important to data analysis:

1. Logarithmic and invert logarithmic transformations
2. Binary partitioning
3. Propensity score matching
4. Mean and hot deck imputations
5. Precision assessments of diagnostic tests
6. Robust variabilities

These methods are, generally, difficult on a statistical software program and easy on a pocket calculator. We should add that pocket calculators work faster, because summary statistics are used. Also you better understand what you are doing. Pocket calculators are wonderful: they enable you to test instantly without the need to download a statistical software program.

The methods can also help you make use of methodologies for which there is little software, like Bhattacharya modeling, fuzzy models, Markov models, binary partitioning, etc.

We do hope that *Statistical Analysis of Clinical Data on a Pocket Calculator 1* and *2* will enhance your understanding and carrying out of medical statistics and help you dig deeper into the fascinating world of statistical data analysis. We recommend to those completing the current books to study, as a next step, the two books entitled *SPSS for Starters 1* and *2* from the same authors.

Lyon, France
March 2012

Ton J. Cleophas
Aeilko H. Zwinderman

# Preface to Second Edition

We as authors were, initially, pretty unsure, whether, in the current era of computer analyses, a work based on pocket calculator analyses of clinical data would be appreciated by medical and health professionals and students. However, within the first two years of publication, over thirty thousand e-copies were sold. From readers' comments we came to realize that statistical software programs had been experienced as black box programs producing lots of p-values, but little answers to scientific questions, and that many readers had not been happy with that situation.

The pocket calculator analyses appeared to be, particularly, appreciated, because they enabled readers for the first time to understand the scientific methods of statistical reasoning and hypothesis testing. So much so that it started something like a new dimension in their professional world.

The reason for a rewrite was to give updated and upgraded versions of the forty chapters from the first editions, including the valuable comments of readers. Like in the textbook complementary to the current work, entitled *SPSS for Starters and 2nd Levelers* (Springer Heidelberg 2015, from the same authors), an improved structure of the chapters was produced, including background, main purpose, scientific question, schematic overview of data files, and reference sections. In addition, for the analysis of more complex data, twenty novel chapters were written. We will show that, also here, a pocket calculator can be very helpful.

For convenience the chapters have been reclassified according to the most basic difference in data characteristics: continuous outcome data (34 chapters) and binary outcome data (26 chapters). Both hypothesized and real data examples are used to explain the sixty pocket calculator methods described. The arithmetic is of a no-more-than high-school level.

Lyon, France
October 2015

Ton J. Cleophas
Aeilko H. Zwinderman

# Contents

# Part I
# Continuous Outcome Data

# Chapter 1
# Data Spread, Standard Deviations

## 1 General Purpose

Repeated measures have a central tendency (averages, two averages, regression lines), and a tendency to depart from the expected central values. In order to estimate the magnitude of the departures from the averages an index is needed. Why not simply add-up departures? However, this does not work, because generally the values higher and lower than the averages tend to even out, and the results would be zero. A pragmatic solution was taken by statisticians around the world. They decided to square the departures first, and then add-up. The add-up sum of the squared departures is called the variance. The square root of the variance is called the standard deviation. This chapter shows how pocket calculators can be used for computation of standard deviations.

## 2 Schematic Overview of Type of Data File

| Outcome |
| --- |
| . |
| . |
| . |
| . |
| . |
| . |
| . |
| . |
| . |
| |
| |



T.J. Cleophas, A.H. Zwinderman, *Clinical Data Analysis on a Pocket Calculator*,
DOI 10.1007/978-3-319-27104-0_1

## 3 Primary Scientific Question

Is the standard deviation an adequate index for spread in the data?

## 4 Data Example

Standard deviations (SDs) are often used for summarizing the spread of the data from a sample. If the spread in the data is small, then the same will be true for the standard deviation. Underneath the calculation is illustrated with the help of a data example.

55
54
51
55
53
53
54
52+

Mean …. => …/8 = 53.375

SD=

55 $(55-53.375)^2$
54 $(54-53.375)^2$
51 $(51-53.375)^2$
55 $(55-53.375)^2$
53 $(53-53.375)^2$
53 $(53-53.375)^2$
54 $(54-53.375)^2$
52 $(52-53.375)^2$+

SD= ………… => ….. / (n-1) => √….=> 1.407885953

## 5 Calculate Standard Deviations

Each scientific pocket calculator has a mode for data-analysis. It is helpful to calculate in a few minutes the mean and standard deviation of a sample.

Calculate standard deviation: mean = 53.375 SD = 1.407885953

The next steps are required:

```
Casio fx-825 scientific
On ... .mode....shift....AC....55....M+....54....M+....51....M+....55....
M+....53....M+....53.....M+....54....M+....52....M+....shift...[x] ....shift....
σxn-1
----------------------------------------------------------------------
Texas TI-30 scientific
On....55....Σ+....54....Σ+....51....Σ+....55....     Σ+....53....Σ+....53....Σ
+....54.... Σ+....52....Σ+....2nd....x....2nd....σxn-1
----------------------------------------------------------------------
Sigma AK 222 and Commodoor
On ... 2ndf ... on ... 55 ... M+ ... 54 ... M+ ... 51 ... M+ ... 55 ... M+ ...  53 ... M+
... 53 ... M+ ... 54 ... M+ ... 52 ... M+ ... x=>M ... MR
----------------------------------------------------------------------
Calculator: Electronic Calculator
On....mode....2....55....M+....54....M+....51....M+....55....M+....53....M
+....53....M+....54....M+....52....M+....Shift....S-var....1....=....
(mean)....Shift....S-var.... 3....(sd)
```

# 6 Conclusion

Repeated measures have a central tendency and tendency to depart from the expected central values. In order to estimate the magnitude of the departures from the averages an index is needed. The add-up sum of the squared departures is used for the purpose, and is called the variance. The square root of the variance is called the standard deviation. This chapter shows how pocket calculators can be used for computation of standard deviations. Sometimes, data files are skewed, and mean values do not mean too much. Instead the modus (the frequentest value) or the median (the value in the middle) are more meaningful (see Chap. 27).

Example:

What is the mean value, what is the SD?

5
4
5
4
5
4
5
4

## 7 Note

More background, theoretical and mathematical information of means, variances, standard deviations, and standard errors (of the mean) is given in Statistics applied to clinical studies 5th edition, Chap. 1, Springer Heidelberg Germany, 2012, from the same authors.

## Chapter 2
# Data Summaries, Histograms, Wide and Narrow Gaussian Curves

## 1 General Purpose

In order to summarize continuous data, either histograms can be plotted or Gaussian curves can be drawn. This chapter is to assess how to summarize your data with the help of a pocket calculator.

## 2 Schematic Overview of Type of Data File

| Outcome |
| --- |
| . |
| . |
| . |
| . |
| . |
| . |
| . |
| . |
| . |

## 3 Primary Scientific Question

How can histograms, otherwise called frequency distributions, and wide and narrow Gaussian curves be used for summarizing continuous data?



T.J. Cleophas, A.H. Zwinderman, *Clinical Data Analysis on a Pocket Calculator*,
DOI 10.1007/978-3-319-27104-0_2

## 4 Hypothesized Data Example

Based on the same data, but with different meaning. The wide one summarizes the data of our trial. The narrow one summarizes the mean of many trials similar to our trial. It can be mathematically proven that this is so.

Continuous data can be plotted in the form of a histogram (upper graph). The upper graph is assumed to present individual cholesterol reductions after one week drug treatment in 1000 patients. The bar in the middle is observed most frequently. The bars on either side grow gradually shorter. The graph, thus, pretty well exhibits two main characteristics of your data namely the place of the mean and the distribution of the individual data against the mean. On the x-axis, frequently called z-axis in statistics, it has individual data. On the y-axis it has "how often". This graph adequately represents the data. It is, however, not adequate for statistical analyses. The lower graph pictures a Gaussian curve, otherwise called normal (distribution) curve. On the x-axis we have, again, the individual data, expressed either in absolute data or in SDs (standard deviations) distant from the mean. See Chap. 1 for calculation information of SDs. On the y-axis the bars have been replaced with a continuous line. It is now impossible to determine from the graph how many patients had a particular outcome. Instead, important inferences can be made. For example, the total area under the curve (AUC) represents:

100 % of the data,
AUC left from mean represents 50 % of the data,
left from −1 SDs it has 15.9 % of the data,
left from −2 SDs it has 2.5 % of the data,
between +2 SDs and −2 SDs we have 95 % of the data
(the 95 % confidence interval of the data).

It is remarkable that the patterns of Gaussian curves from biological data are countless, but that all of them, nonetheless, display the above percentages. This is something like a present from heaven, and it enables to make use of the Gaussian curves for making predictions from your data about future data. However, in order to statistically test your data, we will have to go one step further.

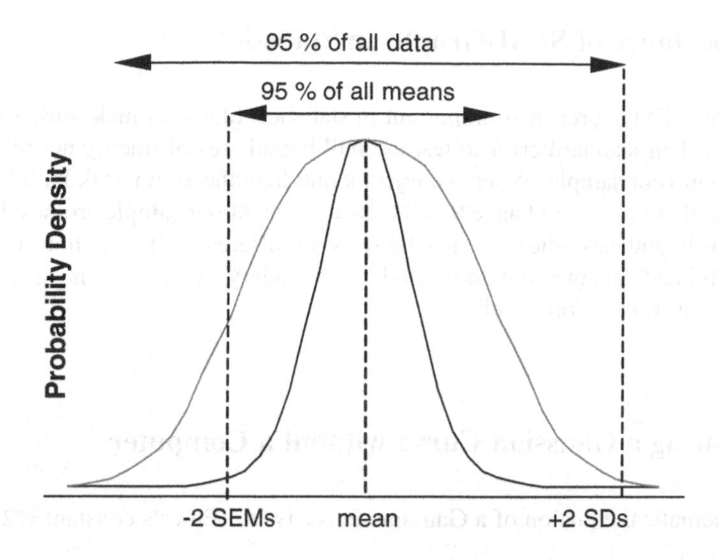


The figure underneath gives two Gaussian curves, a narrow and a wide one. Both are based on the same data, but with different meaning. The wide one summarizes the data of our trial. The narrow one summarizes the mean of many trials similar to our trial. It can be mathematically proven that this is so. However this is beyond scope of the current text. Still, it is easy to conceive that the distribution of all means of many similar trials is narrower and has fewer outliers than the distribution of the actual data from our trial, and that it will center around the mean of our trial, because our trial is assumed to be representative for the entire population. Now why should the mean of many trials be equal to the mean of our trial. The truth is, we have no certainty, but neither do we have any argument to have the overall mean on the left of right side of the measured mean of our data. You may find it hard to believe, but the narrow curve with standard errors of the mean (SEMs), or, simply, SEs on the x-axis can be effectively used for testing important statistical hypotheses, like

1. no difference between new and standard treatment,
2. a real difference,
3. the new treatment is better than the standard treatment,
4. the two treatments are equivalent.

thus, mean $\pm$ 2 SDs (or more precisely 1.96 SDs) represents 95 % of the AUC of the wide distribution, otherwise called the 95 % confidence interval of the data, which means that 95 % of the data of the sample are within. The SEM-curve (narrow one) is narrower than the SD-curve (wide one), because SEM $=$ SD$/\sqrt{n}$ with n $=$ sample size. Mean $\pm$ 2 SEMs (or more precisely 1.96 SEMs) represents 95 % of the means of many trials similar to our trial.

$$\text{SEM} = \text{SD}/\sqrt{n}$$

## 5 Importance of SEM-Graphs in Statistics

Why is this SEM approach so important in statistics. Statistics makes use of mean values and their standard error to test the null hypotheses of finding no difference from zero in your sample. When we reject a null hypothesis at $p < 0.05$, it literally means that there is <5 % chance that the mean value of our sample crosses the area of the null hypothesis where we say there is no difference. It does not mean that many individual data may not go beyond that boundary. Actually, it is just a matter of agreement. But it works well.

## 6 Drawing a Gaussian Curve without a Computer

The mathematical equation of a Gaussian curve is (e = Euler's constant = 2.718)

$y = e^{-1/2\ (x\text{\^{}}2)}$
x = (individual values) / (standard deviation)
x^2 = $x^2$
at $x = 0 \rightarrow y = 1$
at $x = 1 \rightarrow y = 0.607$
at $x = 2 \rightarrow y = 0.135$
at $x = 3 \rightarrow y = 0.011$

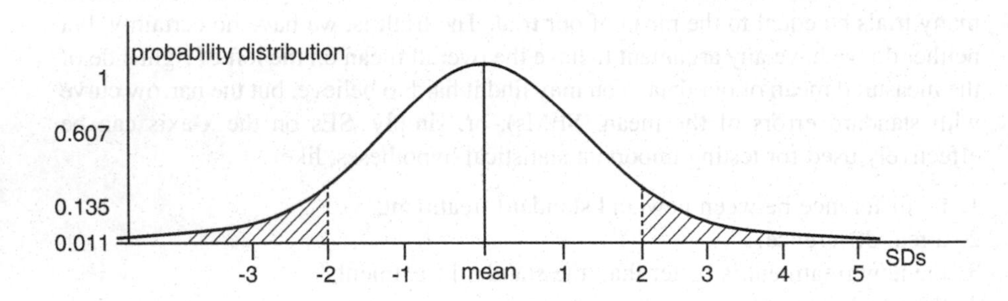


With the help of the above equation, various y-values can, thus, be computed, and in this way a standard Gaussian curve can be drawn.

## 7 Conclusion

In order to summarize continuous data, either histograms can be plotted or Gaussian curves can be drawn. Gaussian curves can be drawn with the help of the Gaussian curve equation $y = e^{-1/2\ (x\ ^2)}$. The above procedure is only entirely correct with larger samples like 100 or so. With small samples data tend to produce somewhat larger spread, and normal distributions turn into t-distributions (see the Chap. 8). But as a first step, before any analysis, histograms and Gaussian curves are convenient even with small samples.

## 8 Note

More background, theoretical and mathematical information of histograms and Gaussian curves is given in Statistics applied to clinical studies 5th edition, Chap. 1, Springer Heidelberg Germany, 2012, from the same authors.

# Chapter 3
# Null-Hypothesis Testing with Graphs

## 1 General Purpose

Because biological processes are *full* of variations, statistics will give no certainties only chances. What chances? Chances that hypotheses are true/untrue. What hypotheses? For example:

1. our mean effect is not different from a 0 effect,
2. it is really different from a 0 effect,
3. it is worse than a 0 effect,

where 0 effect means that your new treatment or any other intervention doesnot work. Statistics is about estimating such chances/testing such hypotheses. Please note that trials often calculate differences between a test treatment and a control treatment, and, subsequently, test whether this difference is larger than 0. A simple way to reduce a study of two groups of data, and, thus, two means to a single mean and single distribution of data, is to take the difference between the two means and compare it with 0.

In the Chap.2 we explained that the data of a trial can be described in the form of a normal distribution graph with SEMs on the x-axis, and that this method is adequate for testing various statistical hypotheses. We will now focus on a very important hypothesis, the null-hypothesis. What it literally means is: no difference from a 0 effect: the mean value of our sample is not different from the value 0. We will try and make a graph of this null-hypothesis, and then assess whether our result is significantly different from the null-hypothesis.



T.J. Cleophas, A.H. Zwinderman, *Clinical Data Analysis on a Pocket Calculator*,
DOI 10.1007/978-3-319-27104-0_3

## 2 Schematic Overview of Type of Data File

| Outcome |
| --- |
| . |
| . |
| . |
| . |
| . |
| . |
| . |
| . |
| . |

## 3 Primary Scientific Question

Is the result of our study significant different from the null-hypothesis?

## 4 Data Example

Let us assume that 1000 patients will be treated with a cholesterol lowering agent. After one week of treatment all cholesterol reductions are summarized with a mean reduction and its standard error. In order to make a graph of our result as compared to the null-hypothesis, our results have to be standardized first.

Mean ± Standard Error

is divided by its own standard error

Mean/Standard Error ± Standard Error/Standard Error =
Mean/Standard Error ± 1

The unit of the standardized results is not mmol/l anymore, but something else. We will call it SEM units (otherwise often called z-values or t-values).

In our example the result is given:

Mean ± Standard Error =
3 ± 1 SEM units

The mathematical equation of the Gaussian curve of our data is obtained from the equation (see also Chap. 2, e = Euler's constant = 2.718).

$$y = e^{-1/2(x^2)}$$

| x = Mean/Standard Error |
|---|
| $x^2 = x^2$ |
| at x = 0 → y = 1 |
| at x = 1 → y = 0.607 |
| at x = 2 → y = 0.135 |
| at x = 3 → y = 0.011 |

The above values can be plotted, and the underneath Gaussian curve H0 (hypothesis 0) will be obtained. It is, actually, a graph of the null-hypothesis.

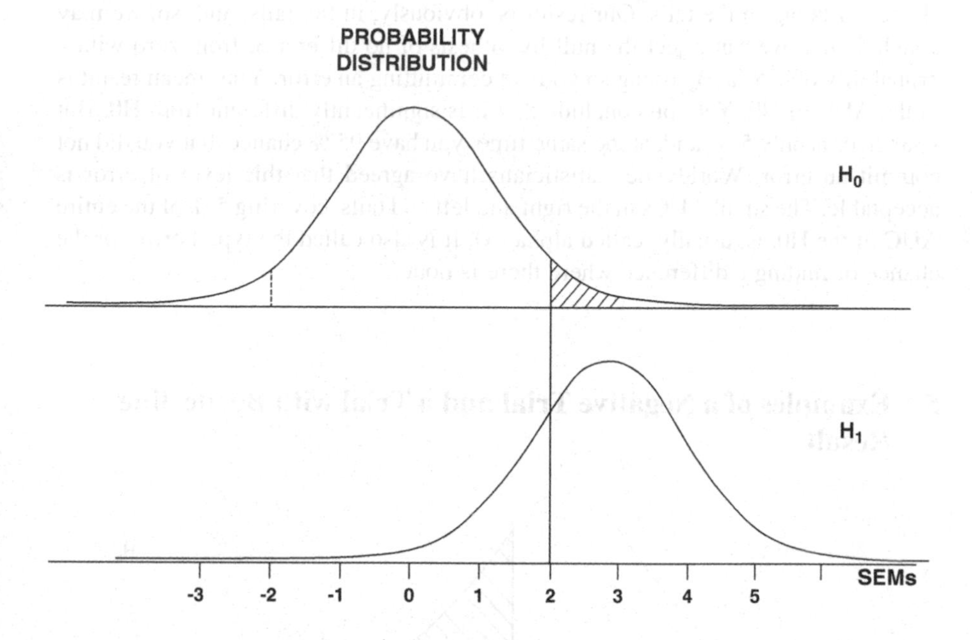


It is now easy to, subsequently, draw a curve of the null-hypothesis H1 (hypothesis 1). It is equally wide and high as the H0 curve, but with an overall mean value of 3 instead of 0. This null-hypothesis is now used to test whether your result is significantly different from zero, meaning that your cholesterol treatment is efficaceous. The following statistical reasoning is used.

As explained in the Chap.2, the Gaussian curve H1 with SEMs on the x-axis is not only a kind of summary of your data, it is also the summary of many trials similar to our trial.

H1 = graph based on the data of our trial with SEMs on the x-axis.
H0 = the same graph with mean 0.

Now we will make a giant leap from our data to the entire population (we can do so, because our data are representative).

H1 = also the summary of many trials similar to ours.
H0 = summary of many trials similar to ours but with an overall mean effect of zero.

We can't prove anything, but we can calculate chances/probabilities. A mean result of 3 is far distant from 0: suppose it belongs to H0. Only 5 % of the H0 trials are >2 SEMs distant from 0. Our mean result is, indeed, > 2 SEMs distant from 0, namely 3 SEMs. This means, that the chance that our result belongs to the H0 trials is <5 %, because the AUC (area under the curve) >2 SEMs and <−2 SEMs is only 5 % of the entire AUC (Chap. 2). We conclude that we have < 5 % of finding this result, and decide that we will reject this small chance.

Any study result larger than 2 SEMs or smaller than – 2 SEMs is in the small tails of the H0 curve, and, if your treatment does not work, you will have less than 5 % chance of being in the tails. Our result is, obviously, in the tails, and, so, we may conclude that we can reject the null-hypothesis of no difference from zero with a probability of < 5 %. By doing so you are committing an error. Your mean result is in the AUC of H0. Yet you conclude that it is significantly different from H0. But your error is only 5 %, and, at the same time, you have 95 % chance that you did not commit an error. Worldwide statisticians have agreed that this level of error is acceptable. The small AUCs in the right and left end tails, covering 5 % of the entire AUC of the H0, is, usually, called alpha (α). It is also called the type I error, or the chance of finding a difference where there is none.

## 5 Examples of a Negative Trial and a Trial with Borderline Result

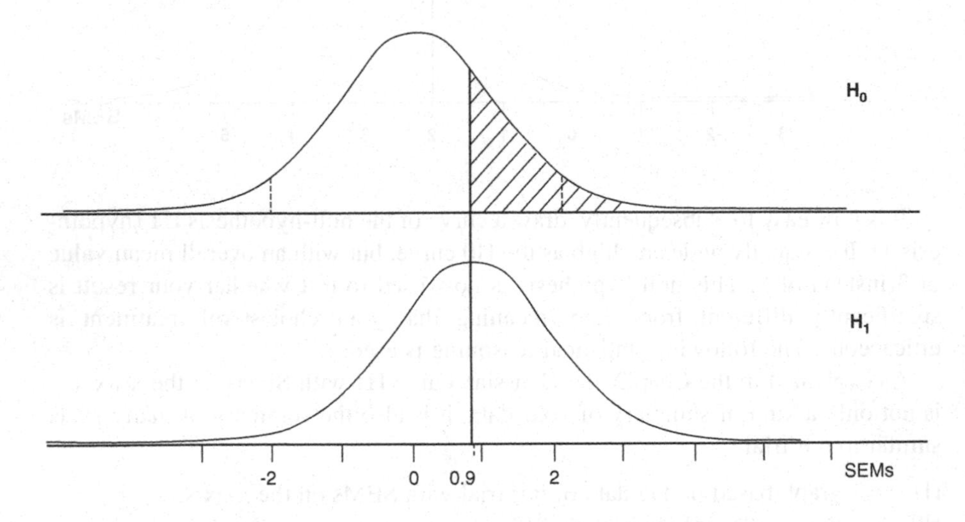


In the above graph the mean of the trial is 0.9 SEMs distant from 0. This result is not on right side of 2 SEMs. The null-hypothesis H0 can, therefore, not be rejected. The AUC (area under the curve) right from 0.9 = not 5 %, but rather 35 % or so of the entire AUC. And so p = 0.35 (35 %).

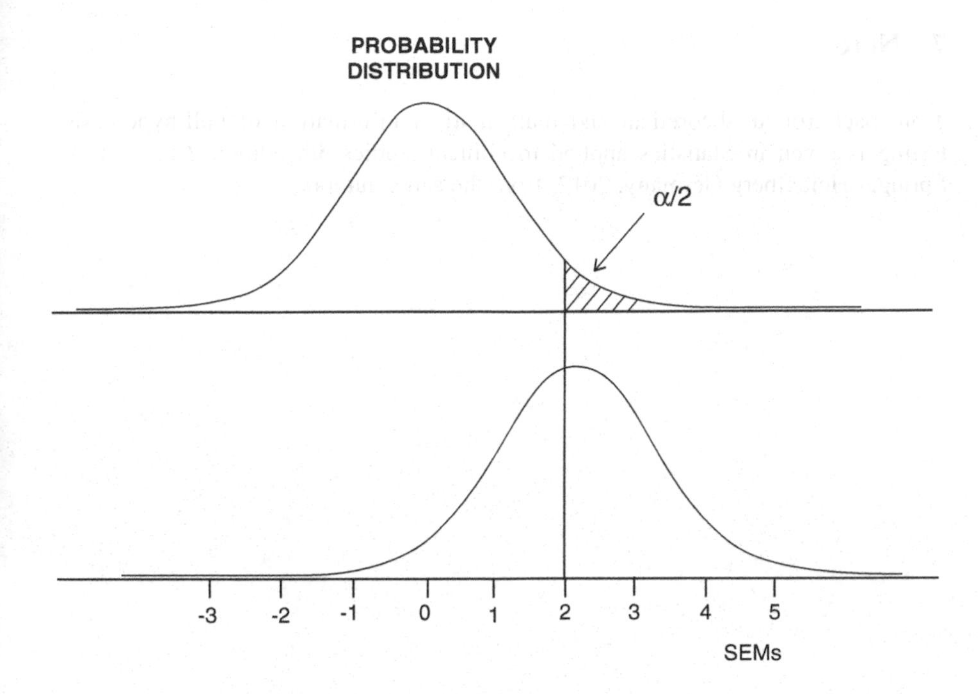


In the above graph the mean is 2 SEMs distant from 0. Alpha level of rejection = 2. The AUC right from 2 is only 5 % of the entire AUC. We will reject H0 at p = 0.05 or 5 %. A p-value of 5 % is, however, a borderline result. We will have, at least, 5 % chance that our conclusion is untrue, and 50 % chance of a type II error (see Chap. 11).

# 6 Conclusion

Because biological processes are *full* of variations, statistics will give no certainties, only chances. What chances? Chances that hypotheses are true/untrue. What hypotheses?: e.g.:

our mean effect is not different from a 0 effect,
a test treatment and a control treatment differ more than zero.

This chapter shows that it is pretty easy to draw a curve of your data and the corresponding null-hypothesis H0. This null-hypothesis can, then, be used to test, whether your result is significantly different from zero, meaning that, e.g., your cholesterol treatment was efficaceous.

# 7 Note

More background, theoretical and mathematical information of null-hypothesis testing is given in Statistics applied to clinical studies 5th edition, Chaps. 1–3, Springer Heidelberg Germany, 2012, from the same authors.

# Chapter 4
# Null-Hypothesis Testing with the T-Table

## 1 General Purpose

In the previous chapter we discussed that the patterns of Gaussian curve from biological data have a constant frequency distribution and that this phenomenon is used for making predictions from your data to future data. However, this is only entirely true with large samples like samples > 100. In practice many studies involve rather small samples and in order for your data from small samples to adequately fit a theoretical frequency distribution we have to replace the Gaussian normal distribution with multiple Gaussian t-distributions which are a little bit wider.

In the twenties the USA government employed wives of workless citizens in the Work Project Administration US. With the help of Monte Carlo Models, and a pocket calculator not yet available at that time, the numerical characteristics of the best fit Gaussian curves for any sample size was calculated. These characteristics were summarized in the famous t-table, which is still the basis of any statistical software program. This chapter is to assess how the t-table can be used for null-hypothesis testing.



T.J. Cleophas, A.H. Zwinderman, *Clinical Data Analysis on a Pocket Calculator*,
DOI 10.1007/978-3-319-27104-0_4

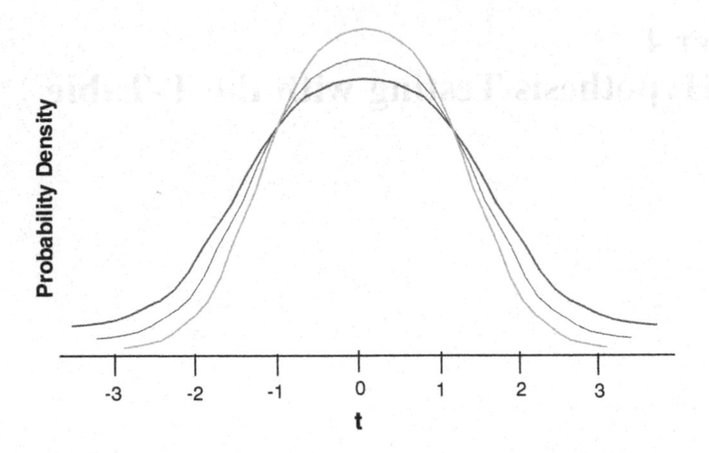


The above graph shows members of the family of t-distributions: with n = 5 the distribution is wide, with n = 10 and n = 100 this is increasingly less so. This chapter is to show how the t-table can be used for testing null-hypotheses of no effects in your data.

## 2 Schematic Overview of Type of Data File

| Outcome |
| --- |
| . |
| . |
| . |
| . |
| . |
| . |
| . |
| . |
| . |

## 3 Primary Scientific Question

Is the result of our study significant different from the null-hypothesis? At what level of statistical significance?

## 4 Data Example

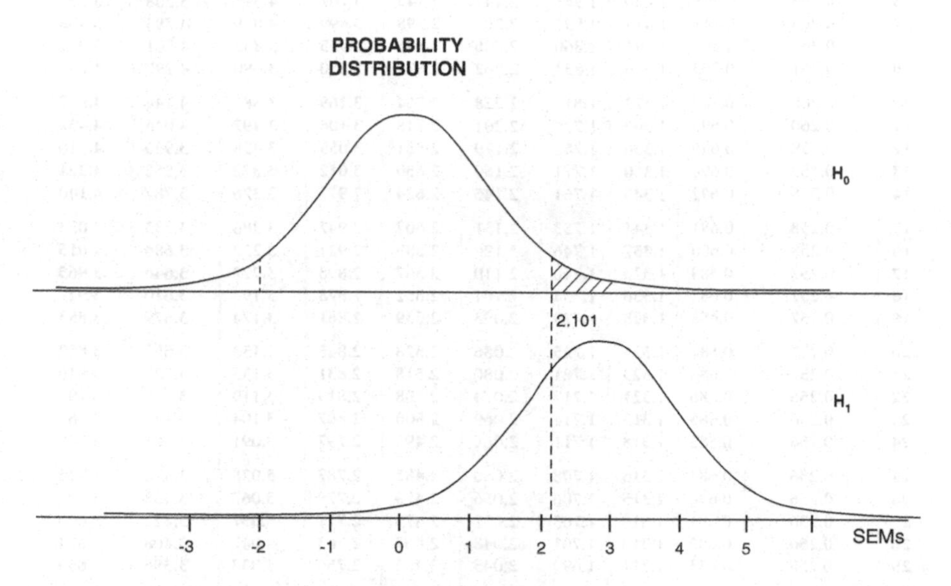


The above graph gives an example of experimental data is given with sample size (n) = 19 and mean result = 2.9 SEMs (SEM-units), and a t-distributed instead of normal frequency distribution. Any result larger than approximately 2 SEMs is statistically significantly different from 0 SEMs at $p < 0.05$. However, our mean of 2.9 SEMs is pretty far away in the <5 % tail, and, so, the chance of being so far away may be a lot smaller than 5 %. Our p-values may be a lot smaller than 0.05. The t-table can be more precise regarding the level of significance, and is given underneath.

## 5 T-Table

| df | One-Tail = .4 | .25 | .1 | .05 | .025 | .01 | .005 | .0025 | .001 | .0005 |
|---|---|---|---|---|---|---|---|---|---|---|
| | Two-Tail = .8 | .5 | .2 | .1 | .05 | .02 | .01 | .005 | .002 | .001 |
| 1 | 0.325 | 1.000 | 3.078 | 6.314 | 12.706 | 31.821 | 63.657 | 127.32 | 318.31 | 636.62 |
| 2 | 0.289 | 0.816 | 1.886 | 2.920 | 4.303 | 6.965 | 9.925 | 14.089 | 22.327 | 31.598 |
| 3 | 0.277 | 0.765 | 1.638 | 2.353 | 3.182 | 4.541 | 5.841 | 7.453 | 10.214 | 12.924 |
| 4 | 0.271 | 0.741 | 1.533 | 2.132 | 2.776 | 3.747 | 4.604 | 5.598 | 7.173 | 8.610 |
| 5 | 0.267 | 0.727 | 1.476 | 2.015 | 2.571 | 3.365 | 4.032 | 4.773 | 5.893 | 6.869 |
| 6 | 0.265 | 0.718 | 1.440 | 1.943 | 2.447 | 3.143 | 3.707 | 4.317 | 5.208 | 5.959 |
| 7 | 0.263 | 0.711 | 1.415 | 1.895 | 2.365 | 2.998 | 3.499 | 4.029 | 4.785 | 5.408 |
| 8 | 0.262 | 0.706 | 1.397 | 1.860 | 2.306 | 2.896 | 3.355 | 3.833 | 4.501 | 5.041 |
| 9 | 0.261 | 0.703 | 1.383 | 1.833 | 2.262 | 2.821 | 3.250 | 3.690 | 4.297 | 4.781 |
| 10 | 0.260 | 0.700 | 1.372 | 1.812 | 2.228 | 2.764 | 3.169 | 3.581 | 4.144 | 4.587 |
| 11 | 0.260 | 0.697 | 1.363 | 1.796 | 2.201 | 2.718 | 3.106 | 3.497 | 4.025 | 4.437 |
| 12 | 0.259 | 0.695 | 1.356 | 1.782 | 2.179 | 2.681 | 3.055 | 3.428 | 3.930 | 4.318 |
| 13 | 0.259 | 0.694 | 1.350 | 1.771 | 2.160 | 2.650 | 3.012 | 3.372 | 3.852 | 4.221 |
| 14 | 0.258 | 0.692 | 1.345 | 1.761 | 2.145 | 2.624 | 2.977 | 3.326 | 3.787 | 4.140 |
| 15 | 0.258 | 0.691 | 1.341 | 1.753 | 2.131 | 2.602 | 2.947 | 3.286 | 3.733 | 4.073 |
| 16 | 0.258 | 0.690 | 1.337 | 1.746 | 2.120 | 2.583 | 2.921 | 3.252 | 3.686 | 4.015 |
| 17 | 0.257 | 0.689 | 1.333 | 1.740 | 2.110 | 2.567 | 2.898 | 3.222 | 3.646 | 3.965 |
| 18 | 0.257 | 0.688 | 1.330 | 1.734 | 2.101 | 2.552 | 2.878 | 3.197 | 3.610 | 3.922 |
| 19 | 0.257 | 0.688 | 1.328 | 1.729 | 2.093 | 2.539 | 2.861 | 3.174 | 3.579 | 3.883 |
| 20 | 0.257 | 0.687 | 1.325 | 1.725 | 2.086 | 2.528 | 2.845 | 3.153 | 3.552 | 3.850 |
| 21 | 0.257 | 0.686 | 1.323 | 1.721 | 2.080 | 2.518 | 2.831 | 3.135 | 3.527 | 3.819 |
| 22 | 0.256 | 0.686 | 1.321 | 1.717 | 2.074 | 2.508 | 2.819 | 3.119 | 3.505 | 3.792 |
| 23 | 0.256 | 0.685 | 1.319 | 1.714 | 2.069 | 2.500 | 2.807 | 3.104 | 3.485 | 3.767 |
| 24 | 0.256 | 0.685 | 1.318 | 1.711 | 2.064 | 2.492 | 2.797 | 3.091 | 3.467 | 3.745 |
| 25 | 0.256 | 0.684 | 1.316 | 1.708 | 2.060 | 2.485 | 2.787 | 3.078 | 3.450 | 3.725 |
| 26 | 0.256 | 0.684 | 1.315 | 1.706 | 2.056 | 2.479 | 2.779 | 3.067 | 3.435 | 3.707 |
| 27 | 0.256 | 0.684 | 1.314 | 1.703 | 2.052 | 2.473 | 2.771 | 3.057 | 3.421 | 3.690 |
| 28 | 0.256 | 0.683 | 1.313 | 1.701 | 2.048 | 2.467 | 2.763 | 3.047 | 3.408 | 3.674 |
| 29 | 0.256 | 0.683 | 1.311 | 1.699 | 2.045 | 2.462 | 2.756 | 3.038 | 3.396 | 3.659 |
| 30 | 0.256 | 0.683 | 1.310 | 1.697 | 2.042 | 2.457 | 2.750 | 3.030 | 3.385 | 3.646 |
| 40 | 0.255 | 0.681 | 1.303 | 1.684 | 2.021 | 2.423 | 2.704 | 2.971 | 3.307 | 3.551 |
| 60 | 0.254 | 0.679 | 1.296 | 1.671 | 2.000 | 2.390 | 2.660 | 2.915 | 3.232 | 3.460 |
| 120 | 0.254 | 0.677 | 1.289 | 1.658 | 1.980 | 2.358 | 2.617 | 2.860 | 3.160 | 3.373 |
| ∞ | 0.253 | 0.674 | 1.282 | 1.645 | 1.960 | 2.326 | 2.576 | 2.807 | 3.090 | 3.291 |

The t-table has a left-end column giving degrees of freedom (≈sample sizes), and two top rows with p-values (areas under the curve = p-values), one-tail meaning that only one end of the curve, two-tail meaning that both ends are assessed simultaneously. The t-table is, furthermore, full of t-values, that, with ∞ degrees of freedom, are equal to z-values (Chap. 36). The t-values are to be understood as mean results of studies, but not expressed in mmol/l, kilograms, but in so-called SEM-units (Standard error of the mean units), that are obtained by dividing your mean result by its own standard error. With many degrees of freedom (large samples) the curve will be a little bit narrower, and more in agreement with nature.

A t-value of 2.9 with 18° of freedom (19 patients and 1 group means we have 19–1 = 18° of freedom) indicates that we will need the row no. 18 of the table. The t-value 2.9 is left from 3.197 and right from 2.878. Now look right up to the second of the two upper rows: we are right from 0.01 and left from 0.005. The p-value equals <0.01.

## 6 Conclusion

In the previous chapter we discussed that the patterns of Gaussian curves from biological data have a constant frequency distribution and that this phenomenon is used for making predictions from your data to future data. However, this is only entirely true with large samples, like samples >100. In practice, many studies involve rather small samples, and in order for your data from small samples to adequately fit a theoretical frequency distribution we have to replace the Gaussian normal distribution with multiple Gaussian-like t-distributions which are a little bit wider. The t-table summarizes the numerical characteristics of these frequency distributions of any size. It helps us to predict precisely the chance and level of a statistically significant effect. E.g., a p value $< 0.002$ means that we have less than 0.2 % of finding such a result if there would be no effect in your data. In clinical term this indicates that your treatment was very efficaceous, and that the chance that this conclusion is erroneous will be less than 0.2 %, at least, if your null-hypothesis is true.

## 7 Note

More background, theoretical and mathematical information of null-hypothesis testing is given Statistics applied to clinical studies 5th edition, Chaps. 1–3, Springer Heidelberg Germany, 2012, from the same authors.

# Chapter 5
# One-Sample Continuous Data (One-Sample *T*-Test, One-Sample Wilcoxon Test)

## 1 General Purpose

Studies where a single outcome per patient is compared to zero may be analyzed with the one-sample *t*-test (see also Chap. 4). The *t*-test is only adequate, if the data can be assumed to follow a Gaussian-like frequency distribution. For non-Gaussian-like data the one-sample Wilcoxon test will be appropriate.

## 2 Schematic Overview of Type of Data File

| Outcome |
|---|
| . |
| . |
| . |
| . |
| . |
| . |
| . |
| . |
| . |
| . |



T.J. Cleophas, A.H. Zwinderman, *Clinical Data Analysis on a Pocket Calculator*,
DOI 10.1007/978-3-319-27104-0_5

## 3 Primary Scientific Question

Are the one-sample *t*-test and one-sample Wilcoxon test adequate for testing whether the result of a one-sample study is significantly different from a zero result?

## 4 Data Example, One-Sample *T*-test

In 10 patients the mean blood pressure reduction after treatment is calculated with the accompanying p-value. A p-value $< 0.05$ indicates, that there is less than 5 % probability that such a decrease will be observed purely by the play of chance. There is, thus, $> 95$ % chance that the decrease is the result of a real blood pressure lowering effect of the treatment. We call such a decrease statistically significant.

| Patient number | mm Hg decrease |
|---|---|
| 1 | 3 |
| 2 | 4 |
| 3 | −2 |
| 4 | 3 |
| 5 | 1 |
| 6 | −3 |
| 7 | 4 |
| 8 | 3 |
| 9 | 2 |
| 10 | 1 |

Is this decrease statistically significant?

Mean decrease = 1.6 mm Hg
SD = 2.4129 mm Hg

From the standard deviation (SD) the standard error (SE) can be calculated using the equation

$$SE = SD/\sqrt{n} \ (n = \text{sample size})$$
$$SE = 2.4129/\sqrt{10} = 0.7636$$

De t-value (t) is the test-statistic of the *t*-test, and is calculated as follows:

$$t = 1.6 \, / \, 0.7636 = 2.095$$

Because the sample size is 10, the test has here $10–1 = 9°$ of freedom.

The t-table underneath shows that with 9° of freedom the t-value should be $> 2.262$ in order to obtain a two-tail result significantly different from zero at $p < 0.05$. With a t-value of 2.095 the level equals: $0.05 < p < 0.10$. This result is close to statistically significant, and is called a trend to significance.

# 5 T-Table

The underneath t-table has a left-end column giving degrees of freedom ($\approx$ sample sizes), and two top rows with p-values (areas under the curve = p-values), one-tail meaning that only one end of the curve, two-tail meaning that both ends are assessed simulataneously. The t-table is, furthermore, full of t-values, that, with $\infty$ degrees of freedom, are equal to z-values (Chap.36). The t-values are to be understood as mean results of studies, but not expressed in mmol/l, kilograms, but in so-called SEM-units (Standard error of the mean units), that are obtained by dividing your mean result by its own standard error. With many degrees of freedom (large samples) the curve will be a little bit narrower, and more in agreement with nature.

| df | One-Tail = .4 | .25 | .1 | .05 | .025 | .01 | .005 | .0025 | .001 | .0005 |
|---|---|---|---|---|---|---|---|---|---|---|
| | Two-Tail = .8 | .5 | .2 | .1 | .05 | .02 | .01 | .005 | .002 | .001 |
| 1 | 0.325 | 1.000 | 3.078 | 6.314 | 12.706 | 31.821 | 63.657 | 127.32 | 318.31 | 636.62 |
| 2 | 0.289 | 0.816 | 1.886 | 2.920 | 4.303 | 6.965 | 9.925 | 14.089 | 22.327 | 31.598 |
| 3 | 0.277 | 0.765 | 1.638 | 2.353 | 3.182 | 4.541 | 5.841 | 7.453 | 10.214 | 12.924 |
| 4 | 0.271 | 0.741 | 1.533 | 2.132 | 2.776 | 3.747 | 4.604 | 5.598 | 7.173 | 8.610 |
| 5 | 0.267 | 0.727 | 1.476 | 2.015 | 2.571 | 3.365 | 4.032 | 4.773 | 5.893 | 6.869 |
| 6 | 0.265 | 0.718 | 1.440 | 1.943 | 2.447 | 3.143 | 3.707 | 4.317 | 5.208 | 5.959 |
| 7 | 0.263 | 0.711 | 1.415 | 1.895 | 2.365 | 2.998 | 3.499 | 4.029 | 4.785 | 5.408 |
| 8 | 0.262 | 0.706 | 1.397 | 1.860 | 2.306 | 2.896 | 3.355 | 3.833 | 4.501 | 5.041 |
| 9 | 0.261 | 0.703 | 1.383 | 1.833 | 2.262 | 2.821 | 3.250 | 3.690 | 4.297 | 4.781 |
| 10 | 0.260 | 0.700 | 1.372 | 1.812 | 2.228 | 2.764 | 3.169 | 3.581 | 4.144 | 4.587 |
| 11 | 0.260 | 0.697 | 1.363 | 1.796 | 2.201 | 2.718 | 3.106 | 3.497 | 4.025 | 4.437 |
| 12 | 0.259 | 0.695 | 1.356 | 1.782 | 2.179 | 2.681 | 3.055 | 3.428 | 3.930 | 4.318 |
| 13 | 0.259 | 0.694 | 1.350 | 1.771 | 2.160 | 2.650 | 3.012 | 3.372 | 3.852 | 4.221 |
| 14 | 0.258 | 0.692 | 1.345 | 1.761 | 2.145 | 2.624 | 2.977 | 3.326 | 3.787 | 4.140 |
| 15 | 0.258 | 0.691 | 1.341 | 1.753 | 2.131 | 2.602 | 2.947 | 3.286 | 3.733 | 4.073 |
| 16 | 0.258 | 0.690 | 1.337 | 1.746 | 2.120 | 2.583 | 2.921 | 3.252 | 3.686 | 4.015 |
| 17 | 0.257 | 0.689 | 1.333 | 1.740 | 2.110 | 2.567 | 2.898 | 3.222 | 3.646 | 3.965 |
| 18 | 0.257 | 0.688 | 1.330 | 1.734 | 2.101 | 2.552 | 2.878 | 3.197 | 3.610 | 3.922 |
| 19 | 0.257 | 0.688 | 1.328 | 1.729 | 2.093 | 2.539 | 2.861 | 3.174 | 3.579 | 3.883 |
| 20 | 0.257 | 0.687 | 1.325 | 1.725 | 2.086 | 2.528 | 2.845 | 3.153 | 3.552 | 3.850 |
| 21 | 0.257 | 0.686 | 1.323 | 1.721 | 2.080 | 2.518 | 2.831 | 3.135 | 3.527 | 3.819 |
| 22 | 0.256 | 0.686 | 1.321 | 1.717 | 2.074 | 2.508 | 2.819 | 3.119 | 3.505 | 3.792 |
| 23 | 0.256 | 0.685 | 1.319 | 1.714 | 2.069 | 2.500 | 2.807 | 3.104 | 3.485 | 3.767 |
| 24 | 0.256 | 0.685 | 1.318 | 1.711 | 2.064 | 2.492 | 2.797 | 3.091 | 3.467 | 3.745 |
| 25 | 0.256 | 0.684 | 1.316 | 1.708 | 2.060 | 2.485 | 2.787 | 3.078 | 3.450 | 3.725 |
| 26 | 0.256 | 0.684 | 1.315 | 1.706 | 2.056 | 2.479 | 2.779 | 3.067 | 3.435 | 3.707 |
| 27 | 0.256 | 0.684 | 1.314 | 1.703 | 2.052 | 2.473 | 2.771 | 3.057 | 3.421 | 3.690 |
| 28 | 0.256 | 0.683 | 1.313 | 1.701 | 2.048 | 2.467 | 2.763 | 3.047 | 3.408 | 3.674 |
| 29 | 0.256 | 0.683 | 1.311 | 1.699 | 2.045 | 2.462 | 2.756 | 3.038 | 3.396 | 3.659 |
| 30 | 0.256 | 0.683 | 1.310 | 1.697 | 2.042 | 2.457 | 2.750 | 3.030 | 3.385 | 3.646 |
| 40 | 0.255 | 0.681 | 1.303 | 1.684 | 2.021 | 2.423 | 2.704 | 2.971 | 3.307 | 3.551 |
| 60 | 0.254 | 0.679 | 1.296 | 1.671 | 2.000 | 2.390 | 2.660 | 2.915 | 3.232 | 3.460 |
| 120 | 0.254 | 0.677 | 1.289 | 1.658 | 1.980 | 2.358 | 2.617 | 2.860 | 3.160 | 3.373 |
| $\infty$ | 0.253 | 0.674 | 1.282 | 1.645 | 1.960 | 2.326 | 2.576 | 2.807 | 3.090 | 3.291 |

# 6 Data Example, One-Sample Wilcoxon Test

| Patient number | mm Hg | Decrease | Smaller one of two add-up sums |
|---|---|---|---|
| 1 | 3 | 6.5 | |
| 2 | 4 | 9.5 | |
| 3 | −2 | 3.5 | 3.5 |
| 4 | 3 | 6.5 | |
| 5 | 1 | 1.5 | |
| 6 | −3 | 6.5 | 6.5 |
| 7 | 4 | 9.5 | |
| 8 | 3 | 6.5 | |
| 9 | 2 | 3.5 | |
| 10 | 1 | 1.5 | |

The example of the previous section will be applied once more for a Wilcoxon analysis. We will first put the differences from zero in ascending order. The patients 5 and 10 are equal (1 mm Hg different from zero), we will give them rank number 1.5 instead of 1 and 2. Patient 3 and 9 are equal (have equal distances from zero), and will be given rank number 3.5 instead of 3 and 4.

The patients 1, 4, 6, and 8 are again equal and will be given rank number 6.5 instead of 5, 6, 7, 8. When each patient has been given an appropriate rank number, all of the positive and all of the negative rank numbers will be added up, and the smaller number of the two will be used for estimating the level of statistical significance. Our add-up sum of negative outcome values is the smaller number, and adds the values of the patients 3 and 6, and equals $3.5 + 6.5 = 10$. According to the underneath Wilcoxon table with 10 number of paires the add-up value of 10 indicates that our p-value equals $< 0.10$. This result is very similar to the result of the above *t*-test. Again a trend to significance is observed at $0.05 < P < 0.10$.

Wilcoxon Test Table

| Number of pairs | P < 0.10 | P < 0.05 | P < 0.01 |
|---|---|---|---|
| **7** | 3 | 2 | 0 |
| **8** | 5 | 2 | 0 |
| **9** | 8 | 6 | 2 |
| **10** | 10 | 8 | 3 |
| **11** | 13 | 11 | 5 |
| **12** | 17 | 14 | 7 |
| **13** | 20 | 17 | 10 |
| **14** | 25 | 21 | 13 |
| **15** | 30 | 25 | 16 |
| **16** | 35 | 30 | 19 |

The first column in the above Wilcoxon test table gives the numbers of patients in your file. Rank numbers of positive and negative differences from zero are separately added up. The second, third, and fourth columns give the smaller one of the two add-up sums required for statistical significance of increasing levels.

## 7 Conclusion

Studies where one outcome in one patient is compared with zero can be analyzed with the one-sample *t*-test. The *t*-test is adequate, if the data can be assumed to follow a Gaussian-like frequency distribution. For non-Gaussian-like data the one-sample Wilcoxon test is appropriate. The example given shows that levels of statistical significance of the two tests are very similar.

## 8 Note

More background, theoretical and mathematical information of testing null-hypothesis testing with t-tests and Wilcoxon tests is given in Statistics applied to clinical studies 5th edition, Chap. 1, Springer Heidelberg Germany, 2012, from the same authors.

# Chapter 6
# Paired Continuous Data (Paired T-Test, Wilcoxon Signed Rank Test)

## 1 General Purpose

Studies where two outcomes in one patient are compared with one another, are often called crossover studies, and the observations are called paired observations.

As paired observations are usually more similar than unpaired observations, special tests are required in order to adjust for a positive correlation between the paired observations.

## 2 Schematic Overview of Type of Data File

| Outcome 1 | Outcome 2 |
|---|---|
| . | . |
| . | . |
| . | . |
| . | . |
| . | . |
| . | . |
| . | . |
| . | . |
| . | . |

 
T.J. Cleophas, A.H. Zwinderman, *Clinical Data Analysis on a Pocket Calculator*,
DOI 10.1007/978-3-319-27104-0_6

## 3 Primary Scientific Question

Is the first outcome significantly different from second one.

## 4 Data Example

The underneath study assesses whether some sleeping pill is more efficacious than a placebo. The hours of sleep is the outcome value.

| Patient number | Outcome 1 | Outcome 2 | Individual Differences |
|---|---|---|---|
| 1 | 6.0 | 5.1 | 0.9 |
| 2 | 7.1 | 8.0 | −0.9 |
| 3 | 8.1 | 3.8 | 4.3 |
| 4 | 7.5 | 4.7 | 3.1 |
| 5 | 6.4 | 5.2 | 1.2 |
| 6 | 7.9 | 5.4 | 2.5 |
| 7 | 6.8 | 4.3 | 2.5 |
| 8 | 6.6 | 6.0 | 0.6 |
| 9 | 7.3 | 3.7 | 3.8 |
| 10 | 5.6 | 6.2 | −0.6 |

Outcome = hours of sleep after treatment

## 5 Analysis: Paired T-Test

Rows may be more convenient than columns, if you use a pocket calculator, because you read the data like you read the lines of a textbook. Two rows of observations in 10 persons are given underneath:

Observations 1:
6.0, 7.1, 8.1, 7.5, 6.4, 7.9, 6.8, 6.6, 7.3, 5.6

Observations 2:
5.1, 8.0, 3.8, 4.4, 5.2, 5.4, 4.3, 6.0, 3.7, 6.2

Individual differences:
0.9, −0.9, 4.3, 3.1, 1.2, 2.5, 2.5, 0.6, 3.8, −0.6

A. not significant
B. $0.05 < p < 0.10$
C. $P < 0.05$
D. $P < 0.01$

Is there a significant difference between the observations 1 and 2, and which level of significance is correct (P = p-value, SD = standard deviation) ?

Mean difference = 1.59
SDs are calculated as demonstrated in the Chap. 1.
SD of mean difference = $\sqrt{(SD_1^2 + SD_2^2)}$
= 1.789 (SD = standard deviation)
SE = SD/√10 = 0.567 (SE = standard error)
t = 1.59/0.567 = 2.81

We have here (10–1) = 9 degrees of freedom, because we have 10 patients and 1 group of patients. According to the underneath t-table the p-value equals $< 0.05$, and we can conclude that a significant difference between the two observations is in the data: the values of row 1 are significantly higher than those of row 2. The answer C is correct.

# 6 T-Table

The t-table has a left-end column giving degrees of freedom (≈ sample sizes), and two top rows with p-values (areas under the curve = p-values), one-tail meaning that only one end of the curve, two-tail meaning that both ends are assessed simultaneously. The t-table is, furthermore, full of t-values, that, with ∞ degrees of freedom, are equal to z-values (Chap. 36). The t-values are to be understood as mean results of studies, but not expressed in mmol/l, kilograms, but in so-called SEM-units (Standard error of the mean units), that are obtained by dividing your mean result by its own standard error. With many degrees of freedom (large samples) the curve will be a little bit narrower, and more in agreement with nature.

| df | One-Tail = .4 | .25 | .1 | .05 | .025 | .01 | .005 | .0025 | .001 | .0005 |
|---|---|---|---|---|---|---|---|---|---|---|
| | Two-Tail = .8 | .5 | .2 | .1 | .05 | .02 | .01 | .005 | .002 | .001 |
| 1 | 0.325 | 1.000 | 3.078 | 6.314 | 12.706 | 31.821 | 63.657 | 127.32 | 318.31 | 636.62 |
| 2 | 0.289 | 0.816 | 1.886 | 2.920 | 4.303 | 6.965 | 9.925 | 14.089 | 22.327 | 31.598 |
| 3 | 0.277 | 0.765 | 1.638 | 2.353 | 3.182 | 4.541 | 5.841 | 7.453 | 10.214 | 12.924 |
| 4 | 0.271 | 0.741 | 1.533 | 2.132 | 2.776 | 3.747 | 4.604 | 5.598 | 7.173 | 8.610 |
| 5 | 0.267 | 0.727 | 1.476 | 2.015 | 2.571 | 3.365 | 4.032 | 4.773 | 5.893 | 6.869 |
| 6 | 0.265 | 0.718 | 1.440 | 1.943 | 2.447 | 3.143 | 3.707 | 4.317 | 5.208 | 5.959 |
| 7 | 0.263 | 0.711 | 1.415 | 1.895 | 2.365 | 2.998 | 3.499 | 4.029 | 4.785 | 5.408 |
| 8 | 0.262 | 0.706 | 1.397 | 1.860 | 2.306 | 2.896 | 3.355 | 3.833 | 4.501 | 5.041 |
| 9 | 0.261 | 0.703 | 1.383 | 1.833 | 2.262 | 2.821 | 3.250 | 3.690 | 4.297 | 4.781 |
| 10 | 0.260 | 0.700 | 1.372 | 1.812 | 2.228 | 2.764 | 3.169 | 3.581 | 4.144 | 4.587 |
| 11 | 0.260 | 0.697 | 1.363 | 1.796 | 2.201 | 2.718 | 3.106 | 3.497 | 4.025 | 4.437 |
| 12 | 0.259 | 0.695 | 1.356 | 1.782 | 2.179 | 2.681 | 3.055 | 3.428 | 3.930 | 4.318 |
| 13 | 0.259 | 0.694 | 1.350 | 1.771 | 2.160 | 2.650 | 3.012 | 3.372 | 3.852 | 4.221 |
| 14 | 0.258 | 0.692 | 1.345 | 1.761 | 2.145 | 2.624 | 2.977 | 3.326 | 3.787 | 4.140 |
| 15 | 0.258 | 0.691 | 1.341 | 1.753 | 2.131 | 2.602 | 2.947 | 3.286 | 3.733 | 4.073 |
| 16 | 0.258 | 0.690 | 1.337 | 1.746 | 2.120 | 2.583 | 2.921 | 3.252 | 3.686 | 4.015 |
| 17 | 0.257 | 0.689 | 1.333 | 1.740 | 2.110 | 2.567 | 2.898 | 3.222 | 3.646 | 3.965 |
| 18 | 0.257 | 0.688 | 1.330 | 1.734 | 2.101 | 2.552 | 2.878 | 3.197 | 3.610 | 3.922 |
| 19 | 0.257 | 0.688 | 1.328 | 1.729 | 2.093 | 2.539 | 2.861 | 3.174 | 3.579 | 3.883 |
| 20 | 0.257 | 0.687 | 1.325 | 1.725 | 2.086 | 2.528 | 2.845 | 3.153 | 3.552 | 3.850 |
| 21 | 0.257 | 0.686 | 1.323 | 1.721 | 2.080 | 2.518 | 2.831 | 3.135 | 3.527 | 3.819 |
| 22 | 0.256 | 0.686 | 1.321 | 1.717 | 2.074 | 2.508 | 2.819 | 3.119 | 3.505 | 3.792 |
| 23 | 0.256 | 0.685 | 1.319 | 1.714 | 2.069 | 2.500 | 2.807 | 3.104 | 3.485 | 3.767 |
| 24 | 0.256 | 0.685 | 1.318 | 1.711 | 2.064 | 2.492 | 2.797 | 3.091 | 3.467 | 3.745 |
| 25 | 0.256 | 0.684 | 1.316 | 1.708 | 2.060 | 2.485 | 2.787 | 3.078 | 3.450 | 3.725 |
| 26 | 0.256 | 0.684 | 1.315 | 1.706 | 2.056 | 2.479 | 2.779 | 3.067 | 3.435 | 3.707 |
| 27 | 0.256 | 0.684 | 1.314 | 1.703 | 2.052 | 2.473 | 2.771 | 3.057 | 3.421 | 3.690 |
| 28 | 0.256 | 0.683 | 1.313 | 1.701 | 2.048 | 2.467 | 2.763 | 3.047 | 3.408 | 3.674 |
| 29 | 0.256 | 0.683 | 1.311 | 1.699 | 2.045 | 2.462 | 2.756 | 3.038 | 3.396 | 3.659 |
| 30 | 0.256 | 0.683 | 1.310 | 1.697 | 2.042 | 2.457 | 2.750 | 3.030 | 3.385 | 3.646 |
| 40 | 0.255 | 0.681 | 1.303 | 1.684 | 2.021 | 2.423 | 2.704 | 2.971 | 3.307 | 3.551 |
| 60 | 0.254 | 0.679 | 1.296 | 1.671 | 2.000 | 2.390 | 2.660 | 2.915 | 3.232 | 3.460 |
| 120 | 0.254 | 0.677 | 1.289 | 1.658 | 1.980 | 2.358 | 2.617 | 2.860 | 3.160 | 3.373 |
| ∞ | 0.253 | 0.674 | 1.282 | 1.645 | 1.960 | 2.326 | 2.576 | 2.807 | 3.090 | 3.291 |

A t-value of 2.81 with 9 degrees of freedom indicates, that we will need the 9th row of the t-values. The upper row of the table gives the area under the curve of the Gaussian-like t-distribution. The t-value 2.81 is left from 3.250, and right from 2.262. Now look right up at the upper row: we are right from 0.05. The p-value is $< 0.05$.

# 7 Alternative Analysis: Wilcoxon Signed Rank Test

The t-tests as reviewed in the previous section are suitable for studies with Gaussian-like data distributions. However, if there are outliers, then the t-tests will not be adequately sensitive, and non-parametric tests will have to be applied. We should add that non-parametric tests are also adequate for testing normally distributed data. And, so, these tests are, actually, universal, and are, therefore, absolutely recommended.

Calculate the p-value using the Wilcoxon signed rank test.

Observations 1:
6.0, 7.1, 8.1, 7.5, 6.4, 7.9, 6.8, 6.6, 7.3, 5.6

Observations 2:
5.1, 8.0, 3.8, 4.4, 5.2, 5.4, 4.3, 6.0, 3.7, 6.2

Individual differences:
0.9, −0.9, 4.3, 3.1, 1.2, 2.5, 2.5, 0.6, 3.6, −0.6

Rank numbers:
3.5, 3.5, 10, 7, 5, 8, 6, 2, 9, 1

A. not significant
B. $0.05 < p < 0.10$
C. $p < 0.05$
D. $p < 0.01$

Is there a significant difference between observations 1 and 2? Which significance level is correct?

The individual differences are given a rank number dependent on their magnitude of difference. If two differences are identical, and if they have for example the rank numbers 3 and 4, then an average rank number is given to both of them, which means 3.5 and 3.5. Next, all positive and all negative rank numbers have to be added up separately. We will find 4.5 and 50.5. According to the Wilcoxon table underneath, with 10 numbers of pairs, the smaller one of the two add-up numbers must be smaller than 8 in order to be able to speak of a p-value $< 0.05$. This is true in our example.

# 8 Wilcoxon Test Table

Wilcoxon Test Table

| Number of pairs | $P < 0.05$ | $P < 0.01$ |
|---|---|---|
| **7** | 2 | 0 |
| | | |
| **8** | 2 | 0 |
| **9** | 6 | 2 |
| **10** | 8 | 3 |
| **11** | 11 | 5 |
| **12** | 14 | 7 |
| **13** | 17 | 10 |
| **14** | 21 | 13 |
| **15** | 25 | 16 |
| **16** | 30 | 19 |

The first column gives the numbers of pairs in your paired data file. Rank numbers of positive and negative differences are separately added up. The second and third columns give the smaller one of the two add-up sums required for statistical significance.

As demonstrated in the above table, also according to the non-parametric Wilcoxon's test the outcome one is significantly larger than the outcome two. The p-value of difference here equals $p < 0.05$. We should add that non-parametric tests take into account more than the t-test, namely, that Non-gaussian-like data are accounted for. If you account more, then you will prove less. That's why the p-value may be somewhat larger.

## 9 Conclusion

The significant effects indicate that the null-hypothesis of no difference between the two outcome can be rejected. The treatment 1 performs better than the treatment 2. It may be prudent to use the non-parametric tests, if normality is doubtful like in the current small data example given. Paired t-tests and Wilcoxon signed rank tests need, just like multivariate data, more than a single outcome variable. However, they cannot assess the effect of predictors on the outcomes, because they do not allow for predictor variables. They can only test the significance of difference between the outcomes.

## 10 Note

The theories of null-hypotheses and frequency distributions and additional examples of paired t-tests and Wilcoxon signed rank tests are reviewed in Statistics applied to clinical studies 5th edition, Chaps. 1 and 2, Springer Heidelberg Germany, 2012, from the same authors.

# Chapter 7
# Unpaired Continuous Data (Unpaired T-Test, Mann-Whitney)

## 1 General Purpose

For the study of two outcomes often two parallel groups of similar age, gender and other characteristics are applied, and the studies are called parallel-group studies, and the two groups are called independent of one another. This study gives examples of parallel-group analyses.

## 2 Schematic Overview of Type of Data File

| Outcome | Parallel-group (1,2) |
|---|---|
| . | 1 |
| . | 1 |
| . | 1 |
| . | 1 |
| . | 1 |
| . | 1 |
| . | 1 |
| . | 1 |
| . | 1 |
| . | 2 |
| . | 2 |
| . | 2 |
| . | 2 |
| . | 2 |
| . | 2 |
| . | 2 |



T.J. Cleophas, A.H. Zwinderman, *Clinical Data Analysis on a Pocket Calculator*,
DOI 10.1007/978-3-319-27104-0_7

## 3 Primary Scientific Question

Are unpaired t-tests or Mann-Whitney tests appropriate for testing, whether the outcome of the first group is significantly different from that of the second group?

## 4 Data Example

The underneath study assesses in 20 patients whether some sleeping pill (parallel-group 1) is more efficacious than a placebo (parallel-group 2). The hours of sleep is the outcome value.

| Outcome parallel-group 1 | Outcome parallel-group 2 |
|---|---|
| 6,0 | 5,1 |
| 7,1 | 8,0 |
| 8,1 | 3,8 |
| 7,5 | 4,7 |
| 6,4 | 5,2 |
| 7,9 | 5,4 |
| 6,8 | 4,3 |
| 6,6 | 6,0 |
| 7,3 | 3,7 |
| 5,6 | 6,2 |

Outcome = hours of sleep after treatment

## 5 Unpaired T-Test

Two age- and gender- matched parallel-groups are compared with one another. For the calculation of SDs see the Chap. 1.

group 1:
6.0, 7.1, 8.1, 7.5, 6.4, 7.9, 6.8, 6.6, 7.3, 5.6

group 2:
5.1, 8.0, 3.8, 4.4, 5.2, 5.4, 4.3, 6.0, 3.7, 6.2

Mean group 1 = 6.93 SD = 0.806 SE = SD/$\sqrt{10}$ = 0.255
Mean Group 2 = 5.21 SD = 1.299 SE = SD/$\sqrt{10}$ = 0.411

A. not significant
B. $0.05 < p < 0.10$
C. $p < 0.05$
D. $p < 0.01$

Is there a significant difference between the two groups, which level of significance is correct?

Mean standard deviation (SD)
6.93 0.806
5.21 - 1.299 pepe
1.72

$$\text{pooled SE} = \sqrt{\left(\frac{0.806^2}{10} + \frac{1.299^2}{10}\right)} = 0.483.$$

The t-value = (6.93 – 5.21) / 0.483 = 3.56.

20–2 = 18 degrees of freedom, because we have 20 patients and 2 groups.

According to the t-table of page 16 the p-value is < 0.01, and we can conclude that that a very significant difference exists between the two groups. The values of group 1 are higher than those of group 2. The answer D is correct.

# 6 T-Table

The t-table has a left-end column giving degrees of freedom (≈ sample sizes), and two top rows with p-values (areas under the curve = p - values), one-tail meaning that only one end of the curve, two-tail meaning that both ends are assessed simultaneously. The t-table is, furthermore, full of t-values, that, with ∞ degrees of freedom, are equal to z-values (Chap.36). The t-values are to be understood as mean results of studies, but not expressed in mmol/l, kilograms, but in so-called SEM-units (Standard error of the mean units), that are obtained by dividing your mean result by its own standard error. With many degrees of freedom (large samples) the curve will be a little bit narrower, and more in agreement with nature.

| df | One-Tail = .4 | .25 | .1 | .05 | .025 | .01 | .005 | .0025 | .001 | .0005 |
|---|---|---|---|---|---|---|---|---|---|---|
| | Two-Tail = .8 | .5 | .2 | .1 | .05 | .02 | .01 | .005 | .002 | .001 |
| 1 | 0.325 | 1.000 | 3.078 | 6.314 | 12.706 | 31.821 | 63.657 | 127.32 | 318.31 | 636.62 |
| 2 | 0.289 | 0.816 | 1.886 | 2.920 | 4.303 | 6.965 | 9.925 | 14.089 | 22.327 | 31.598 |
| 3 | 0.277 | 0.765 | 1.638 | 2.353 | 3.182 | 4.541 | 5.841 | 7.453 | 10.214 | 12.924 |
| 4 | 0.271 | 0.741 | 1.533 | 2.132 | 2.776 | 3.747 | 4.604 | 5.598 | 7.173 | 8.610 |
| 5 | 0.267 | 0.727 | 1.476 | 2.015 | 2.571 | 3.365 | 4.032 | 4.773 | 5.893 | 6.869 |
| 6 | 0.265 | 0.718 | 1.440 | 1.943 | 2.447 | 3.143 | 3.707 | 4.317 | 5.208 | 5.959 |
| 7 | 0.263 | 0.711 | 1.415 | 1.895 | 2.365 | 2.998 | 3.499 | 4.029 | 4.785 | 5.408 |
| 8 | 0.262 | 0.706 | 1.397 | 1.860 | 2.306 | 2.896 | 3.355 | 3.833 | 4.501 | 5.041 |
| 9 | 0.261 | 0.703 | 1.383 | 1.833 | 2.262 | 2.821 | 3.250 | 3.690 | 4.297 | 4.781 |
| 10 | 0.260 | 0.700 | 1.372 | 1.812 | 2.228 | 2.764 | 3.169 | 3.581 | 4.144 | 4.587 |
| 11 | 0.260 | 0.697 | 1.363 | 1.796 | 2.201 | 2.718 | 3.106 | 3.497 | 4.025 | 4.437 |
| 12 | 0.259 | 0.695 | 1.356 | 1.782 | 2.179 | 2.681 | 3.055 | 3.428 | 3.930 | 4.318 |
| 13 | 0.259 | 0.694 | 1.350 | 1.771 | 2.160 | 2.650 | 3.012 | 3.372 | 3.852 | 4.221 |
| 14 | 0.258 | 0.692 | 1.345 | 1.761 | 2.145 | 2.624 | 2.977 | 3.326 | 3.787 | 4.140 |
| 15 | 0.258 | 0.691 | 1.341 | 1.753 | 2.131 | 2.602 | 2.947 | 3.286 | 3.733 | 4.073 |
| 16 | 0.258 | 0.690 | 1.337 | 1.746 | 2.120 | 2.583 | 2.921 | 3.252 | 3.686 | 4.015 |
| 17 | 0.257 | 0.689 | 1.333 | 1.740 | 2.110 | 2.567 | 2.898 | 3.222 | 3.646 | 3.965 |
| 18 | 0.257 | 0.688 | 1.330 | 1.734 | 2.101 | 2.552 | 2.878 | 3.197 | 3.610 | 3.922 |
| 19 | 0.257 | 0.688 | 1.328 | 1.729 | 2.093 | 2.539 | 2.861 | 3.174 | 3.579 | 3.883 |
| 20 | 0.257 | 0.687 | 1.325 | 1.725 | 2.086 | 2.528 | 2.845 | 3.153 | 3.552 | 3.850 |
| 21 | 0.257 | 0.686 | 1.323 | 1.721 | 2.080 | 2.518 | 2.831 | 3.135 | 3.527 | 3.819 |
| 22 | 0.256 | 0.686 | 1.321 | 1.717 | 2.074 | 2.508 | 2.819 | 3.119 | 3.505 | 3.792 |
| 23 | 0.256 | 0.685 | 1.319 | 1.714 | 2.069 | 2.500 | 2.807 | 3.104 | 3.485 | 3.767 |
| 24 | 0.256 | 0.685 | 1.318 | 1.711 | 2.064 | 2.492 | 2.797 | 3.091 | 3.467 | 3.745 |
| 25 | 0.256 | 0.684 | 1.316 | 1.708 | 2.060 | 2.485 | 2.787 | 3.078 | 3.450 | 3.725 |
| 26 | 0.256 | 0.684 | 1.315 | 1.706 | 2.056 | 2.479 | 2.779 | 3.067 | 3.435 | 3.707 |
| 27 | 0.256 | 0.684 | 1.314 | 1.703 | 2.052 | 2.473 | 2.771 | 3.057 | 3.421 | 3.690 |
| 28 | 0.256 | 0.683 | 1.313 | 1.701 | 2.048 | 2.467 | 2.763 | 3.047 | 3.408 | 3.674 |
| 29 | 0.256 | 0.683 | 1.311 | 1.699 | 2.045 | 2.462 | 2.756 | 3.038 | 3.396 | 3.659 |
| 30 | 0.256 | 0.683 | 1.310 | 1.697 | 2.042 | 2.457 | 2.750 | 3.030 | 3.385 | 3.646 |
| 40 | 0.255 | 0.681 | 1.303 | 1.684 | 2.021 | 2.423 | 2.704 | 2.971 | 3.307 | 3.551 |
| 60 | 0.254 | 0.679 | 1.296 | 1.671 | 2.000 | 2.390 | 2.660 | 2.915 | 3.232 | 3.460 |
| 120 | 0.254 | 0.677 | 1.289 | 1.658 | 1.980 | 2.358 | 2.617 | 2.860 | 3.160 | 3.373 |
| ∞ | 0.253 | 0.674 | 1.282 | 1.645 | 1.960 | 2.326 | 2.576 | 2.807 | 3.090 | 3.291 |

A t-value of 3.56 with 18 degrees of freedom indicates, that we will need the 18th row of t-values. The upper row of the table gives the area under the curve of the Gaussian-like t-distribution. The t-value 3.56 is left from 4.297, right from 3.250. Now look right up at the upper row: we are right from 0.01. The p-value equals <0.01. The hours of sleep during the sleeping pill are significantly better than those during placebo.

# 7 Mann-Whitney test

Like the Wilcoxon test, being the non-parametric alternative for the paired t-test, the Mann-Whitney test is the non-parametric alternative for the unpaired t-test. Also this test is applicable for all kinds of data, and, therefore, particularly, to be recommended for investigators with little affection to medical statistics.

Calculate the p-value of the difference between two groups of 10 patients with the help of this test.

group 1:
6.0, 7.1, 8.1, 7.5, 6.4, 7.9, 6.8, 6.6, 7.3, 5.6,

group 2:
5.1, 8.0, 3.8, 4.4, 5.2, 5.4, 4.3, 6.0, 3.7, 6.2

A. not significant
B. $0.05 < p < 0.10$
C. $p < 0.05$
D. $p < 0.01$

Is there a significant difference between the two groups? What significance level is correct?

All values are ranked together in ascending order of magnitude. The values from group 1 are printed thin, those from group 2 are printed fat. Add a rank number to each value. If there are identical values, for example, the rank numbers 9 and 10, then replace those rank numbers with average rank numbers, 9.5 and 9.5. Subsequently, all fat printed rank numbers are added up, and so are the thin printed rank numbers. We will find the values 142.5 for fat print, and 67.5 for thin print.

According to the underneath Mann-Whitney table. the difference should be larger than 71 in order for the significance level of difference to be $<0.05$. We find a difference of 75, which means that there is a p-value $<0.05$ and that the difference between the two groups is, thus, significant.

3.7 1
3.8 2
4.3 3
4.4 4
5.1 5
5.2 6
5.4 7
**5.6 8**
**6.0 9.5**
6.0 9.5
6.2 11
**6.4 12**
**6.6 13**

**6.8** **14**
**7.1** **15**
**7.3** **16**
**7.5** **17**
**7.9** **18**
8.0 19
**8.1** **20**

The Mann-Whitney test tables are given underneath. The values are the minimal differences that are statistically significant with a p-value <0.01 (upper table), and p < 0.05 (lower table). The upper row gives the size of Group 1, the left column the size of Group 2.

# 8 Mann-Whitney Table P < 0.01

P < 0.01 levels

P<0.01 levels

| $n_1 \rightarrow$ / $n_2 \downarrow$ | 2 | 3 | 4 | 5 | 6 | 7 | 8 | 9 | 10 | 11 | 12 | 13 | 14 | 15 |
|---|---|---|---|---|---|---|---|---|---|---|---|---|---|---|
| 4 | | | 10 | | | | | | | | | | | |
| 5 | | 6 | 11 | 17 | | | | | | | | | | |
| 6 | | 7 | 12 | 18 | 26 | | | | | | | | | |
| 7 | | 7 | 13 | 20 | 27 | 36 | | | | | | | | |
| 8 | 3 | 8 | 14 | 21 | 29 | 38 | 49 | | | | | | | |
| 9 | 3 | 8 | 15 | 22 | 31 | 40 | 51 | 63 | | | | | | |
| 10 | 3 | 9 | 15 | 23 | 32 | 42 | 53 | 65 | 78 | | | | | |
| 11 | 4 | 9 | 16 | 24 | 34 | 44 | 55 | 68 | 81 | 96 | | | | |
| 12 | 4 | 10 | 17 | 26 | 35 | 46 | 58 | 71 | 85 | 99 | 115 | | | |
| 13 | 4 | 10 | 18 | 27 | 37 | 48 | 60 | 73 | 88 | 103 | 119 | 137 | | |
| 14 | 4 | 11 | 19 | 28 | 38 | 50 | 63 | 76 | 91 | 106 | 123 | 141 | 160 | |
| 15 | 4 | 11 | 20 | 29 | 40 | 52 | 65 | 79 | 94 | 110 | 127 | 145 | 164 | 185 |
| 16 | 4 | 12 | 21 | 31 | 42 | 54 | 67 | 82 | 97 | 114 | 131 | 150 | 169 | |
| 17 | 5 | 12 | 21 | 32 | 43 | 56 | 70 | 84 | 100 | 117 | 135 | 154 | | |
| 18 | 5 | 13 | 22 | 33 | 45 | 58 | 72 | 87 | 103 | 121 | 139 | | | |
| 19 | 5 | 13 | 23 | 34 | 46 | 60 | 74 | 90 | 107 | 124 | | | | |
| 20 | 5 | 14 | 24 | 35 | 48 | 62 | 77 | 93 | 110 | | | | | |
| 21 | 6 | 14 | 25 | 37 | 50 | 64 | 79 | 95 | | | | | | |
| 22 | 6 | 15 | 26 | 38 | 51 | 66 | 82 | | | | | | | |
| 23 | 6 | 15 | 27 | 39 | 53 | 68 | | | | | | | | |
| 24 | 6 | 16 | 28 | 40 | 55 | | | | | | | | | |
| 25 | 6 | 16 | 28 | 42 | | | | | | | | | | |
| 26 | 7 | 17 | 29 | | | | | | | | | | | |
| 27 | 7 | 17 | | | | | | | | | | | | |
| 28 | 7 | | | | | | | | | | | | | |

## 9 Mann-Whitney Table P < 0.05

P < 0.05 levels

P<0.05 levels

| $n_2$ ↓ \ $n_1$ → | 2 | 3 | 4 | 5 | 6 | 7 | 8 | 9 | 10 | 11 | 12 | 13 | 14 | 15 |
|---|---|---|---|---|---|---|---|---|---|---|---|---|---|---|
| 5 | | | | 15 | | | | | | | | | | |
| 6 | | | 10 | 16 | 23 | | | | | | | | | |
| 7 | | | 10 | 17 | 24 | 32 | | | | | | | | |
| 8 | | | 11 | 17 | 25 | 34 | 43 | | | | | | | |
| 9 | | 6 | 11 | 18 | 26 | 35 | 45 | 56 | | | | | | |
| 10 | | 6 | 12 | 19 | 27 | 37 | 47 | 58 | 71 | | | | | |
| 11 | | 6 | 12 | 20 | 28 | 38 | 49 | 61 | 74 | 87 | | | | |
| 12 | | 7 | 13 | 21 | 30 | 40 | 51 | 63 | 76 | 90 | 106 | | | |
| 13 | | 7 | 14 | 22 | 31 | 41 | 53 | 65 | 79 | 93 | 109 | 125 | | |
| 14 | | 7 | 14 | 22 | 32 | 43 | 54 | 67 | 81 | 96 | 112 | 129 | 147 | |
| 15 | | 8 | 15 | 23 | 33 | 44 | 56 | 70 | 84 | 99 | 115 | 133 | 151 | 171 |
| 16 | | 8 | 15 | 24 | 34 | 46 | 58 | 72 | 86 | 102 | 119 | 137 | 155 | |
| 17 | | 8 | 16 | 25 | 36 | 47 | 60 | 74 | 89 | 105 | 122 | 140 | | |
| 18 | | 8 | 16 | 26 | 37 | 49 | 62 | 76 | 92 | 108 | 125 | | | |
| 19 | 3 | 9 | 17 | 27 | 38 | 50 | 64 | 78 | 94 | 111 | | | | |
| 20 | 3 | 9 | 18 | 28 | 39 | 52 | 66 | 81 | 97 | | | | | |
| 21 | 3 | 9 | 18 | 29 | 40 | 53 | 68 | 83 | | | | | | |
| 22 | 3 | 10 | 19 | 29 | 42 | 55 | 70 | | | | | | | |
| 23 | 3 | 10 | 19 | 30 | 43 | 57 | | | | | | | | |
| 24 | 3 | 10 | 20 | 31 | 44 | | | | | | | | | |
| 25 | 3 | 11 | 20 | 32 | | | | | | | | | | |
| 26 | 3 | 11 | 21 | | | | | | | | | | | |
| 27 | 4 | 11 | | | | | | | | | | | | |
| 28 | 4 | | | | | | | | | | | | | |

## 10 Conclusion

For the study of two outcomes two parallel groups of similar age, gender and other characteristics are often applied, and the studies are called parallel-group studies, and the two groups are called independent of one another. Unpaired tests, like the unpaired t-test and Mann-Whitney test are appropriate for analysis.

## 11 Note

The theories of null-hypotheses and frequency distributions and additional examples of unpaired t-tests and Mann-Whitney tests are reviewed in Statistics applied to clinical studies 5th edition, Chaps. 1 and 2, entitled "Hypotheses data stratification" and "The analysis of efficacy data", Springer Heidelberg Germany, 2012, from the same authors.

# Chapter 8
# Linear Regression (Regression Coefficient, Correlation Coefficient and Their Standard Errors)

## 1 General Purpose

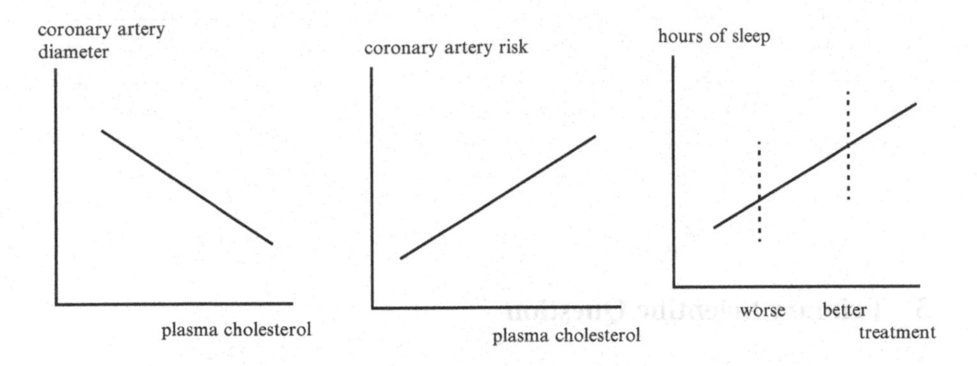


Similarly to unpaired t-tests and Mann-Whitney tests (Chap. 7), linear regression can be used to test whether there is a significant difference between two unpaired treatment modalities. To see how it works, picture the above linear regression of cholesterol levels and diameters of coronary arteries. It shows that the higher the cholesterol, the narrower the coronary arteries. Cholesterol levels are drawn on the x-axis, coronary diameters on the y-axis, and the best fit regression line about the data can be calculated. If coronary artery risk is measured on the y-axis instead of coronary artery diameter, then a positive correlation will be observed (right graph). Instead of a continuous variable on the x-axis, a binary variable can be adequately used, such as two treatment modalities, e.g. a worse and better treatment. With hours of sleep on the y-axis, a nice linear regression analysis can be performed: the better the sleeping treatment, the larger the numbers of sleeping hours. The treatment modality is called the x-variable. Other terms for the x-variable are independent variable, exposure variable, and predictor variable. The hours of sleep is called



T.J. Cleophas, A.H. Zwinderman, *Clinical Data Analysis on a Pocket Calculator*,
DOI 10.1007/978-3-319-27104-0_8

the y-variable, otherwise called the dependent or outcome variable. This chapter is to show how a linear simple linear analysis works.

## 2 Schematic Overview of Type of Data File

| Outcome | binary predictor |
|---|---|
| . | . |
| . | . |
| . | . |
| . | . |
| . | . |
| . | . |
| . | . |
| . | . |
| . | . |
| | |
| | |
| | |
| | |

## 3 Primary Scientific Question

Can linear regression be applied to demonstrate, whether, in an unpaired two group study, one treatment is significantly more efficaceous than the other treatment.

## 4 Data Example

In a parallel-group study of 20 patients 10 are treated with a sleeping pill, 10 with a placebo. The data file is given underneath.

| Outcome | group (1, 2) |
|---|---|
| 6.0 | 1 |
| 7.1 | 1 |
| 8.1 | 1 |
| 7.5 | 1 |
| 6.4 | 1 |
| 7.9 | 1 |
| 6.8 | 1 |

| | |
|---|---|
| 6.6 | 1 |
| 7.3 | 1 |
| 5.6 | 1 |
| 5.1 | 2 |
| 8.0 | 2 |
| 3.8 | 2 |
| 4.7 | 2 |
| 5.2 | 2 |
| 5.4 | 2 |
| 4.3 | 2 |
| 6.0 | 2 |
| 3.7 | 2 |
| 6.2 | 2 |

The group variable has 1 for sleeping pill group, 2 for the placebo.
The outcome variable is hours of sleep after treatment.

# 5 Analysis: Linear Regression

The equation of a linear regression model is given by

$$y = a + bx,$$

with y named the dependent variable and x the independent variable.

The line drawn from this linear function provides the best fit line for the data given, where y = socalled dependent, and x = independent variable, b = regression coefficient, a = intercept:

a and b from the equation y = a+bx can be calculated.

$$b = \text{regression coefficient} = \frac{\sum (x-\bar{x})(y-\bar{y})}{\sum (x-\bar{x})^2}$$

$$a = \text{intercept} = \bar{y}\text{-}b\bar{x}$$

r = correlation coefficient = another important determinant and looks a lot like b.

$$r = \frac{\sum (x-\bar{x})(y-\bar{y})}{\sqrt{\sum (x-\bar{x})^2 \sum (y-\bar{y})^2}}$$

r = measure for the strength of association between the y and x-data. The stronger the association, the better y predicts x, with +1 and -1 as respectively maximal and minimal r-values.

If b and r are statistically significantly larger than 0, then x is a significant predictor of y, and in the example given, this would mean, that there is a significant difference between the groups 1 and 2. One group performs better than the other, and, so, one treatment is better than the other.

## 6 Electronic Calculator for Linear Regression

We will use Electronic Calculator (see Chap. 1) for computations. First, we will calculate the b and r values.

```
Command:
click ON....click MODE....press 3....press 1....press SHIFT, MODE,
and again 1....press =....start entering the data.... [1,   6,0]....[1,
7,1]....[1,   8,1] etc....
```

In order to obtain the b value, press: shift, S-VAR, ►, ►, 2, = .
In order to obtain the r value, press: shift, S-VAR, ►, ►, 3, = .
The b value equals 1.70, the r value equals -0.643.
We wish to assess whether these two values are significantly larger than 0.

The standard error of r $= (1 - r^2) / \sqrt{}(n - 2)$

The r-value is a kind of summary-value of data, and follows a t-distribution, and can, thus, be tested with a *t*-test.

$t = \mid r$ / (its standard error) $\mid$
$t = 0.643 \times 5.539$
$t = 3.56$

This value is much larger than 1.96, and, thus, r is significantly larger / smaller than 0. The t-value of b can be demonstrated to be equally 3.56.

## 7 T-Table

In the above study we have 20 outcome values and 2 groups. According to the underneath t-table, with 20-2 degrees of freedom (see 18th row of t-values), a t-value of 3.56 will be close to 3.610. This means, that the treatment 1 is better than the treatment 0 at a p-value close to 0.002. The t-table is briefly explained in the legends underneath the t-table. It is more fully explained in the Chaps. 4, 5, 6 and 7.

| df | One-Tail = .4<br>Two-Tail = .8 | .25<br>.5 | .1<br>.2 | .05<br>.1 | .025<br>.05 | .01<br>.02 | .005<br>.01 | .0025<br>.005 | .001<br>.002 | .0005<br>.001 |
|---|---|---|---|---|---|---|---|---|---|---|
| 1 | 0.325 | 1.000 | 3.078 | 6.314 | 12.706 | 31.821 | 63.657 | 127.32 | 318.31 | 636.62 |
| 2 | 0.289 | 0.816 | 1.886 | 2.920 | 4.303 | 6.965 | 9.925 | 14.089 | 22.327 | 31.598 |
| 3 | 0.277 | 0.765 | 1.638 | 2.353 | 3.182 | 4.541 | 5.841 | 7.453 | 10.214 | 12.924 |
| 4 | 0.271 | 0.741 | 1.533 | 2.132 | 2.776 | 3.747 | 4.604 | 5.598 | 7.173 | 8.610 |
| 5 | 0.267 | 0.727 | 1.476 | 2.015 | 2.571 | 3.365 | 4.032 | 4.773 | 5.893 | 6.869 |
| 6 | 0.265 | 0.718 | 1.440 | 1.943 | 2.447 | 3.143 | 3.707 | 4.317 | 5.208 | 5.959 |
| 7 | 0.263 | 0.711 | 1.415 | 1.895 | 2.365 | 2.998 | 3.499 | 4.029 | 4.785 | 5.408 |
| 8 | 0.262 | 0.706 | 1.397 | 1.860 | 2.306 | 2.896 | 3.355 | 3.833 | 4.501 | 5.041 |
| 9 | 0.261 | 0.703 | 1.383 | 1.833 | 2.262 | 2.821 | 3.250 | 3.690 | 4.297 | 4.781 |
| 10 | 0.260 | 0.700 | 1.372 | 1.812 | 2.228 | 2.764 | 3.169 | 3.581 | 4.144 | 4.587 |
| 11 | 0.260 | 0.697 | 1.363 | 1.796 | 2.201 | 2.718 | 3.106 | 3.497 | 4.025 | 4.437 |
| 12 | 0.259 | 0.695 | 1.356 | 1.782 | 2.179 | 2.681 | 3.055 | 3.428 | 3.930 | 4.318 |
| 13 | 0.259 | 0.694 | 1.350 | 1.771 | 2.160 | 2.650 | 3.012 | 3.372 | 3.852 | 4.221 |
| 14 | 0.258 | 0.692 | 1.345 | 1.761 | 2.145 | 2.624 | 2.977 | 3.326 | 3.787 | 4.140 |
| 15 | 0.258 | 0.691 | 1.341 | 1.753 | 2.131 | 2.602 | 2.947 | 3.286 | 3.733 | 4.073 |
| 16 | 0.258 | 0.690 | 1.337 | 1.746 | 2.120 | 2.583 | 2.921 | 3.252 | 3.686 | 4.015 |
| 17 | 0.257 | 0.689 | 1.333 | 1.740 | 2.110 | 2.567 | 2.898 | 3.222 | 3.646 | 3.965 |
| 18 | 0.257 | 0.688 | 1.330 | 1.734 | 2.101 | 2.552 | 2.878 | 3.197 | 3.610 | 3.922 |
| 19 | 0.257 | 0.688 | 1.328 | 1.729 | 2.093 | 2.539 | 2.861 | 3.174 | 3.579 | 3.883 |
| 20 | 0.257 | 0.687 | 1.325 | 1.725 | 2.086 | 2.528 | 2.845 | 3.153 | 3.552 | 3.850 |
| 21 | 0.257 | 0.686 | 1.323 | 1.721 | 2.080 | 2.518 | 2.831 | 3.135 | 3.527 | 3.819 |
| 22 | 0.256 | 0.686 | 1.321 | 1.717 | 2.074 | 2.508 | 2.819 | 3.119 | 3.505 | 3.792 |
| 23 | 0.256 | 0.685 | 1.319 | 1.714 | 2.069 | 2.500 | 2.807 | 3.104 | 3.485 | 3.767 |
| 24 | 0.256 | 0.685 | 1.318 | 1.711 | 2.064 | 2.492 | 2.797 | 3.091 | 3.467 | 3.745 |
| 25 | 0.256 | 0.684 | 1.316 | 1.708 | 2.060 | 2.485 | 2.787 | 3.078 | 3.450 | 3.725 |
| 26 | 0.256 | 0.684 | 1.315 | 1.706 | 2.056 | 2.479 | 2.779 | 3.067 | 3.435 | 3.707 |
| 27 | 0.256 | 0.684 | 1.314 | 1.703 | 2.052 | 2.473 | 2.771 | 3.057 | 3.421 | 3.690 |
| 28 | 0.256 | 0.683 | 1.313 | 1.701 | 2.048 | 2.467 | 2.763 | 3.047 | 3.408 | 3.674 |
| 29 | 0.256 | 0.683 | 1.311 | 1.699 | 2.045 | 2.462 | 2.756 | 3.038 | 3.396 | 3.659 |
| 30 | 0.256 | 0.683 | 1.310 | 1.697 | 2.042 | 2.457 | 2.750 | 3.030 | 3.385 | 3.646 |
| 40 | 0.255 | 0.681 | 1.303 | 1.684 | 2.021 | 2.423 | 2.704 | 2.971 | 3.307 | 3.551 |
| 60 | 0.254 | 0.679 | 1.296 | 1.671 | 2.000 | 2.390 | 2.660 | 2.915 | 3.232 | 3.460 |
| 120 | 0.254 | 0.677 | 1.289 | 1.658 | 1.980 | 2.358 | 2.617 | 2.860 | 3.160 | 3.373 |
| ∞ | 0.253 | 0.674 | 1.282 | 1.645 | 1.960 | 2.326 | 2.576 | 2.807 | 3.090 | 3.291 |

The t-table has a left-end column giving degrees of freedom (≈ sample sizes), and two top rows with p-values (areas under the curve = p - values), one-tail meaning that only one end of the curve, two-tail meaning that both ends are assessed simulataneously. The t-table is, furthermore, full of t-values, that, with ∞ degrees of freedom, are equal to z-values (Chap. 36). The t-values are to be understood as mean results of studies, but not expressed in mmol/l, kilograms, but in so-called SEM-units (Standard error of the mean units), that are obtained by dividing your mean result by its own standard error. With many degrees of freedom (large samples) the curve will be a little bit narrower, and more in agreement with nature.

# 8 Conclusion

We can conclude that the correlation and regression coefficients, r and b, are very significant with p-values close to 0.002. This demonstrates that the sleeping scores after active treatment are generally larger than after placebo treatment. The significant correlation between the treatment modality and the numbers of sleeping hours can be interpreted as a significant difference in treatment efficacy of the two treatment modalities. An interesting thing about linear regression is that the linear regression equation can be used for estimating from future x-values the best fit predictions of y-values. In our example we only have two x-values, but if you have more of them, the size of your dependent variable can pretty well be predicted from measured x-values, particularly, if your level of statistical significance is very high (r values close to +1 or -1). R values > 95 % are, actually, applied for validating quantitative diagnostic tests (see also Chap. 25).

# 9 Note

More examples of linear regression analyses are given in Statistics applied to clinical studies 5th edition, Chaps. 14 and 15, Springer Heidelberg Germany, 2012, from the same authors.

# Chapter 9
# Kendall-Tau Regression for Ordinal Data

## 1 General Purpose

Linear regressions (Chaps. 8 and 10) are adequate for outcomes with continuous data, otherwise called scale data. Continuous data have a stepping pattern, where the steps have equal intervals. If, in a regression model, the outcome data have a stepping pattern, but the intervals are not equal, then the term ordinal data is more appropriate for such data, and regression testing of ranks is more appropriate. The data need to be tested according to the magnitude of their rank numbers. This chapter is to assess how rank testing of regression models performs as compared to traditional linear regression.

## 2 Schematic Overview of Type of Data File

| Rank number exposure | Rank number outcome |
|---|---|
| . | . |
| . | . |
| . | . |
| . | . |
| . | . |
| . | . |
| . | . |



T.J. Cleophas, A.H. Zwinderman, *Clinical Data Analysis on a Pocket Calculator*,
DOI 10.1007/978-3-319-27104-0_9

## 3 Primary Scientific Question

Is rank testing of linear by linear data adequately sensitive for testing linear data where the order of data may be more important than the magnitude itself. The latter type of data is usually called ordinal data.

## 4 Data Example

In a short stay hospital the numbers hospitalization days were used to predict the numbers of medical complications. It was assumed, that, the longer the stay, the more risk of multiple complications.

| Patient no. | 1 | 2 | 3 | 4 | 5 | 6 | 7 | 8 |
|---|---|---|---|---|---|---|---|---|
| Days in hospital | 8 | 9 | 10 | 2 | 3 | 4 | 5 | 16 |
| Numbers if complications | 5 | 20 | 16 | 2 | 1 | 3 | 6 | 12 |

A traditional linear regression of the above data will produce a correlation coefficient of 0.695 with a p-value of 0.056, which is not statistically significant.

From the above 8 patients the underneath rank numbers different in magnitude can be obtained:

| Rank numbers of days in hospital | 5 | 6 | 7 | 1 | 2 | 3 | 4 | 8 |
|---|---|---|---|---|---|---|---|---|
| Rank numbers of complications | 4 | 8 | 7 | 2 | 1 | 3 | 5 | 6 |

| Rank numbers of days in hospital in ascending order | 1 | 2 | 3 | 4 | 5 | 6 | 7 | 8 |
|---|---|---|---|---|---|---|---|---|
| Corresponding rank numbers of complications | 2 | 1 | 3 | 5 | 4 | 8 | 7 | 6 |

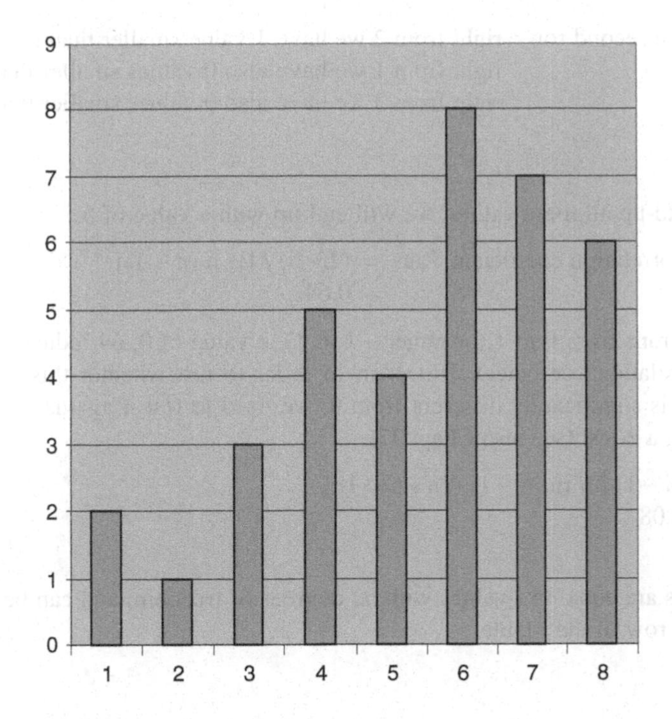


The above graph (with days in hospital on the x-axis and numbers of complications on the y-axis), shows, that, although the numbers of complications tend to increase with the numbers of hospital days their relationship is far from linear. However, the data are not continuous, but ranks, and rank testing is the appropriate analysis.

# 5 Rank Correlation Testing

We wish to know, whether the rank correlation between days in hospital and numbers of complication is statistically significant.

1 2 3 4 5 6 7 8
2 1 3 5 4 8 7 6

In the above second row right from 2 we have 6 values larger than 2,
right from 1 we have also 6 values larger than 1,
right from 3 we have also 5 values larger than 3,
....
....

If we add-up all these values, we will end up with a value of 23.

In the above second row right from 2 we have 1 value smaller than 2,
right from 1 we have also 0 values smaller than 1,
right from 3 we have also 0 values smaller than 3,
....
....

If we add-up all these values, we will end up with a value of 5.

The rank correlation coefficient Tau $= (23–5) / [½ n (n − 1)]$
$= 0.64.$

As Tau runs from 0 to 1 (or rather −1 to 1), a value of 0, 64 indicates a pretty strong correlation coefficient. However, in order to test whether this correlation coefficient is significantly different from 0, we need to test it against its standard error, using a z-test (see also Chap. 37).

$z = (|Tau| - 1) / \sqrt{} [n (n - 1) (2n + 5) / 18]$
$z = 17 / 8.08$
$z = 2.10$

Z-values are equal to t-values with ∞ degrees of freedom, and can be found in the bottom row of the t-table.

# 6 T-Table

Our above z-value, 2.10. is >1.960. According to the underneath t-table (bottom row) a z-value >1.960, corresponding with a two-tail p-value of <0.05 (look right up at the 2nd upper area under the curve row), indicates, that a significant correlation between the x- and y-variable. The days in hospital is closer to numbers of complications than could happen by chance at p-value slightly <0.05, and the association between the two variables is, thus, statistically significant.

| df | One-Tail = .4 | .25 | .1 | .05 | .025 | .01 | .005 | .0025 | .001 | .0005 |
|---|---|---|---|---|---|---|---|---|---|---|
| | Two-Tail = .8 | .5 | .2 | .1 | .05 | .02 | .01 | .005 | .002 | .001 |
| 1 | 0.325 | 1.000 | 3.078 | 6.314 | 12.706 | 31.821 | 63.657 | 127.32 | 318.31 | 636.62 |
| 2 | 0.289 | 0.816 | 1.886 | 2.920 | 4.303 | 6.965 | 9.925 | 14.089 | 22.327 | 31.598 |
| 3 | 0.277 | 0.765 | 1.638 | 2.353 | 3.182 | 4.541 | 5.841 | 7.453 | 10.214 | 12.924 |
| 4 | 0.271 | 0.741 | 1.533 | 2.132 | 2.776 | 3.747 | 4.604 | 5.598 | 7.173 | 8.610 |
| 5 | 0.267 | 0.727 | 1.476 | 2.015 | 2.571 | 3.365 | 4.032 | 4.773 | 5.893 | 6.869 |
| 6 | 0.265 | 0.718 | 1.440 | 1.943 | 2.447 | 3.143 | 3.707 | 4.317 | 5.208 | 5.959 |
| 7 | 0.263 | 0.711 | 1.415 | 1.895 | 2.365 | 2.998 | 3.499 | 4.029 | 4.785 | 5.408 |
| 8 | 0.262 | 0.706 | 1.397 | 1.860 | 2.306 | 2.896 | 3.355 | 3.833 | 4.501 | 5.041 |
| 9 | 0.261 | 0.703 | 1.383 | 1.833 | 2.262 | 2.821 | 3.250 | 3.690 | 4.297 | 4.781 |
| 10 | 0.260 | 0.700 | 1.372 | 1.812 | 2.228 | 2.764 | 3.169 | 3.581 | 4.144 | 4.587 |
| 11 | 0.260 | 0.697 | 1.363 | 1.796 | 2.201 | 2.718 | 3.106 | 3.497 | 4.025 | 4.437 |
| 12 | 0.259 | 0.695 | 1.356 | 1.782 | 2.179 | 2.681 | 3.055 | 3.428 | 3.930 | 4.318 |
| 13 | 0.259 | 0.694 | 1.350 | 1.771 | 2.160 | 2.650 | 3.012 | 3.372 | 3.852 | 4.221 |
| 14 | 0.258 | 0.692 | 1.345 | 1.761 | 2.145 | 2.624 | 2.977 | 3.326 | 3.787 | 4.140 |
| 15 | 0.258 | 0.691 | 1.341 | 1.753 | 2.131 | 2.602 | 2.947 | 3.286 | 3.733 | 4.073 |
| 16 | 0.258 | 0.690 | 1.337 | 1.746 | 2.120 | 2.583 | 2.921 | 3.252 | 3.686 | 4.015 |
| 17 | 0.257 | 0.689 | 1.333 | 1.740 | 2.110 | 2.567 | 2.898 | 3.222 | 3.646 | 3.965 |
| 18 | 0.257 | 0.688 | 1.330 | 1.734 | 2.101 | 2.552 | 2.878 | 3.197 | 3.610 | 3.922 |
| 19 | 0.257 | 0.688 | 1.328 | 1.729 | 2.093 | 2.539 | 2.861 | 3.174 | 3.579 | 3.883 |
| 20 | 0.257 | 0.687 | 1.325 | 1.725 | 2.086 | 2.528 | 2.845 | 3.153 | 3.552 | 3.850 |
| 21 | 0.257 | 0.686 | 1.323 | 1.721 | 2.080 | 2.518 | 2.831 | 3.135 | 3.527 | 3.819 |
| 22 | 0.256 | 0.686 | 1.321 | 1.717 | 2.074 | 2.508 | 2.819 | 3.119 | 3.505 | 3.792 |
| 23 | 0.256 | 0.685 | 1.319 | 1.714 | 2.069 | 2.500 | 2.807 | 3.104 | 3.485 | 3.767 |
| 24 | 0.256 | 0.685 | 1.318 | 1.711 | 2.064 | 2.492 | 2.797 | 3.091 | 3.467 | 3.745 |
| 25 | 0.256 | 0.684 | 1.316 | 1.708 | 2.060 | 2.485 | 2.787 | 3.078 | 3.450 | 3.725 |
| 26 | 0.256 | 0.684 | 1.315 | 1.706 | 2.056 | 2.479 | 2.779 | 3.067 | 3.435 | 3.707 |
| 27 | 0.256 | 0.684 | 1.314 | 1.703 | 2.052 | 2.473 | 2.771 | 3.057 | 3.421 | 3.690 |
| 28 | 0.256 | 0.683 | 1.313 | 1.701 | 2.048 | 2.467 | 2.763 | 3.047 | 3.408 | 3.674 |
| 29 | 0.256 | 0.683 | 1.311 | 1.699 | 2.045 | 2.462 | 2.756 | 3.038 | 3.396 | 3.659 |
| 30 | 0.256 | 0.683 | 1.310 | 1.697 | 2.042 | 2.457 | 2.750 | 3.030 | 3.385 | 3.646 |
| 40 | 0.255 | 0.681 | 1.303 | 1.684 | 2.021 | 2.423 | 2.704 | 2.971 | 3.307 | 3.551 |
| 60 | 0.254 | 0.679 | 1.296 | 1.671 | 2.000 | 2.390 | 2.660 | 2.915 | 3.232 | 3.460 |
| 120 | 0.254 | 0.677 | 1.289 | 1.658 | 1.980 | 2.358 | 2.617 | 2.860 | 3.160 | 3.373 |
| ∞ | 0.253 | 0.674 | 1.282 | 1.645 | 1.960 | 2.326 | 2.576 | 2.807 | 3.090 | 3.291 |

The t-table has a left-end column giving degrees of freedom ($\approx$ sample sizes), and two top rows with p-values (areas under the curve $=$ p $-$ values), one-tail meaning that only one end of the curve, two-tail meaning that both ends are assessed simultaneously. The t-table is, furthermore, full of t-values, that, with $\infty$ degrees of freedom, are equal to z-values (Chap. 36). The t-values are to be understood as mean results of studies, but not expressed in mmol/l, kilograms, but in so-called SEM-units (Standard error of the mean units), that are obtained by dividing your mean result by its own standard error. With many degrees of freedom (large samples) the curve will be a little bit narrower, and more in agreement with nature.

# 7 Conclusion

The Kendall-Tau regression assesses just like the traditional linear regression the level of linear relationships between two variables. However, Kendall-Tau provides a slightly less sensitive result. The traditional linear model (as described in the Chaps. 8 and 10) of the rank data produces an r-value of 0.857 and a p-value of 0.007. Why so? Unlike traditional linear regression, Kendall-Tau takes into account that the intervals between the rank values may not be identical. Also, just like in the non-parametric tests for comparing treatment groups and treatment modalities (Chaps. 6, 7, and 34), if an analysis method takes into account more, it will usually produce less spectacular _results. More in general, if you account more, you will prove less.

# 8 Note

More background, theoretical and mathematical information of rank testing are given in the Chaps 5, 6, and 7 of this work, and in the Chaps. 1, 2, 4, 9, 13, SPSS for starters and 2nd levelers 2nd edition, Springer Heidelberg Germany, 2015, from the same authors.

# Chapter 10
# Paired Continuous Data, Analysis with Help of Correlation Coefficients

## 1 General Purpose

The t-value obtained from an unpaired analysis of paired data produces biased results. This is, because the level of correlation between unpaired data is assumed to be zero, and this may not be true for paired observations. Particularly, repeated measurements in one subject produces usually results more similar than those from single measurements in separate subjects. Repeated measurements, thus, tends to produce a positive correlation. However, this is not always true. Negative correlations will be observed, if completely different treatment effects are examined in one subject. This is, because the responders to one treatment are more at risk of being non-responder to the other treatment and vice versa. Indeed, correlations is a very basic phenomenon in statistical analyses, and it almost entirely determines the results of regression analyses.

This chapter is to examine the performance of correlation coefficients (r-values or R-values) for testing paired data, alternative to the traditional paired t-test and Wilcoxon test. The advantage is, that correlation coefficients unmask, how the level of correlation between repeated measures affect the overall uncertainty in crossover study, and other repeated measures studies.

## 2 Schematic Overview of Type of Data File

| Outcome 1 | Outcome 2 |
|---|---|
| . | . |
| . | . |
| . | . |
| . | . |

(continued)



T.J. Cleophas, A.H. Zwinderman, *Clinical Data Analysis on a Pocket Calculator*,
DOI 10.1007/978-3-319-27104-0_10

| Outcome 1 | Outcome 2 |
|---|---|
| . | . |
| . | . |
| . | . |
| . | . |
| . | . |
| | |
| | |
| | |

## 3 Primary Scientific Question

Can analysis with help of correlation coefficients be used to test in a crossover study whether the first outcome significantly different from second one?

## 4 Data Example

In a crossover study 10 patients are treated with a sleeping pill and a placebo. The first 11 patients of the 20 patient data file is given underneath.

| Outcome 1 | Outcome 2 |
|---|---|
| 6.0 | 5.1 |
| 7.1 | 8.0 |
| 8.1 | 3.8 |
| 7.5 | 4.7 |
| 6.4 | 5.2 |
| 7.9 | 5.4 |
| 6.8 | 4.3 |
| 6.6 | 6.0 |
| 7.3 | 3.7 |
| 5.6 | 6.2 |

Outcome = hours of sleep after treatment

## 5 Unpaired T-Test of Paired Data, the Wrong Analysis

outcome 1:
6.0, 7.1, 8.1, 7.5, 6.4, 7.9, 6.8, 6.6, 7.3, 5.6

outcome 2:
5.1, 8.0, 3.8, 4.4, 5.2, 5.4, 4.3, 6.0, 3.7, 6.2

Mean group 1 = 6.93 SD = 0.806 SE = SD/√10 = 0.255
Mean Group 2 = 5.21 SD = 1.299 SE = SD/√10 = 0.411

Is there a significant difference between the two groups, which level of significance is correct (SD = standard deviation, SE = standard error of the mean)?

Mean standard deviation (SD)

6.93 0.806
5.21 - 1.299

1.72

$$\text{pooled SE} = \sqrt{\left(\frac{0.806^2}{10} + \frac{1.299^2}{10}\right)} = 0.483.$$

The t-value = (6.93 – 5.21) / 0.483 = 3.56.

# 6 Paired T-Test of Paired Data

Observations 1:
6.0, 7.1, 8.1, 7.5, 6.4, 7.9, 6.8, 6.6, 7.3, 5.6

Observations 2:
5.1, 8.0, 3.8, 4.4, 5.2, 5.4, 4.3, 6.0, 3.7, 6.2

Individual differences:
0.9, -0.9, 4.3, 3.1, 1.2, 2.5, 2.5, 0.6, 3.8, -0.6

A. not significant
B. $0.05 < p < 0.10$
C. $P < 0.05$
D. $P < 0.01$

Is there a significant difference between the observations 1 and 2, and which level of significance is correct?

Mean difference = 1.59
SD of mean difference = 1.789 (SD = standard deviation)
SE = SD/√10 = 0.566 (SE = standard error)
t = 1,59 / 0,566 = 2.81

This t-value of 2.81 is a lot smaller than the one from the above unpaired t-test (3.56). Obviously, the correlation between the first and second observations is negative. We will first calculate the level of correlation, r, and then use the underneath 2nd equation for adjustment of the overestimated t-value instead of the first equation.

Standard error unpaired differences $= \frac{\sqrt{(SD_1^2 + SD_2^2)}}{n}$

Standard error paired differences $= \frac{\sqrt{(SD_1^2 + SD_2^2 - 2\, r\, SD_1 . SD_2)}}{n}$

# 7 Linear Regression for Adjustment of Erroneous T-Value from Sect. 5

The equation of a linear regression model is given by

$$y = a + bx,$$

with y named the dependent variable and x the independent variable.

The line drawn from this linear function provides the best fit for the data given, where y = so-called dependent, and x = independent variable, b = regression coefficient, a = intercept:

a and b from the equation $y = a + bx$ can be calculated.

$$b = \text{regression coefficient} = \frac{\sum (x - \overline{x})(y - \overline{y})}{\sum (x - \overline{x})^2}$$

$a = \text{intercept} = \overline{y} - b\overline{x}$

r = correlation coefficient = another important determinant and looks a lot like b.

$$r = \frac{\sum (x - \overline{x})(y - \overline{y})}{\sqrt{\sum (x - \overline{x})^2 \sum (y - \overline{y})^2}}$$

r = measure for the strength of association between y and x-data. The stronger the association, the better y predicts x, with +1 and −1 as maximal and minimal r-values.

We will use the Electronic Calculator (see Chap. 1).

Command:

```
click ON....click MODE....press 3....press 1....press SHIFT, MODE,
and again 1....press=....start entering the data.... [1, 6,0]....[1,
7,1]....[1, 8,1] etc....
In order to obtain the b value, press: shift, S-VAR, ►, ►, 2, = .
In order to obtain the r value, press: shift, S-VAR, ►, ►, 3, = .
```

The b value equals 1.70, the r value equals −0.643.
Standard error paired differences (* is symbol of multiplication)

$$= \frac{\sqrt{\left(SD_1{}^2+SD_2{}^2-2\ r\ SD_1.SD_2\right)}}{\sqrt{n}}$$

$$= \frac{\sqrt{\left(0.806^2+1.299^2+1,286 * 0.806 * 1.299\right)}}{\sqrt{10}}$$

$$= 0.607$$

Adjusted t-value $= 1.72 / 0.607$
$=2.83$

This adjusted t-value is approximately equal to the t-value obtained from the paired t-test in above Sect. 6. The small difference is due to shortening of the final digits during calculations.

## 8 T-Table

According to 9th row of the underneath t-table (10 subjects in one group means 9 degrees of freedom), with $t = 2.83$, the t-value is between 2.821 and 3.250. The corresponding p-values can be observed in the second top row. It is between 0.02 and 0.01, and, thus, $< 0.02$.

The t-table has a left-end column giving degrees of freedom (≈ sample sizes), and two top rows with p-values (areas under the curve = p-values), one-tail meaning that only one end of the curve, two-tail meaning that both ends are assessed simultaneously. The t-table is, furthermore, full of t-values, that, with ∞ degrees of freedom, are equal to z-values (Chap. 36). The t-values are to be understood as mean results of studies, but not expressed in mmol/l, kilograms, but in so-called SEM-units (Standard error of the mean units), that are obtained by dividing your mean result by its own standard error. With many degrees of freedom (large samples) the curve will be a little bit narrower, and more in agreement with nature.

| *df* | One-Tail = .4<br>Two-Tail = .8 | .25<br>.5 | .1<br>.2 | .05<br>.1 | .025<br>.05 | .01<br>.02 | .005<br>.01 | .0025<br>.005 | .001<br>.002 | .0005<br>.001 |
|---|---|---|---|---|---|---|---|---|---|---|
| 1 | 0.325 | 1.000 | 3.078 | 6.314 | 12.706 | 31.821 | 63.657 | 127.32 | 318.31 | 636.62 |
| 2 | 0.289 | 0.816 | 1.886 | 2.920 | 4.303 | 6.965 | 9.925 | 14.089 | 22.327 | 31.598 |
| 3 | 0.277 | 0.765 | 1.638 | 2.353 | 3.182 | 4.541 | 5.841 | 7.453 | 10.214 | 12.924 |
| 4 | 0.271 | 0.741 | 1.533 | 2.132 | 2.776 | 3.747 | 4.604 | 5.598 | 7.173 | 8.610 |
| 5 | 0.267 | 0.727 | 1.476 | 2.015 | 2.571 | 3.365 | 4.032 | 4.773 | 5.893 | 6.869 |
| 6 | 0.265 | 0.718 | 1.440 | 1.943 | 2.447 | 3.143 | 3.707 | 4.317 | 5.208 | 5.959 |
| 7 | 0.263 | 0.711 | 1.415 | 1.895 | 2.365 | 2.998 | 3.499 | 4.029 | 4.785 | 5.408 |
| 8 | 0.262 | 0.706 | 1.397 | 1.860 | 2.306 | 2.896 | 3.355 | 3.833 | 4.501 | 5.041 |
| 9 | 0.261 | 0.703 | 1.383 | 1.833 | 2.262 | 2.821 | 3.250 | 3.690 | 4.297 | 4.781 |
| 10 | 0.260 | 0.700 | 1.372 | 1.812 | 2.228 | 2.764 | 3.169 | 3.581 | 4.144 | 4.587 |
| 11 | 0.260 | 0.697 | 1.363 | 1.796 | 2.201 | 2.718 | 3.106 | 3.497 | 4.025 | 4.437 |
| 12 | 0.259 | 0.695 | 1.356 | 1.782 | 2.179 | 2.681 | 3.055 | 3.428 | 3.930 | 4.318 |
| 13 | 0.259 | 0.694 | 1.350 | 1.771 | 2.160 | 2.650 | 3.012 | 3.372 | 3.852 | 4.221 |
| 14 | 0.258 | 0.692 | 1.345 | 1.761 | 2.145 | 2.624 | 2.977 | 3.326 | 3.787 | 4.140 |
| 15 | 0.258 | 0.691 | 1.341 | 1.753 | 2.131 | 2.602 | 2.947 | 3.286 | 3.733 | 4.073 |
| 16 | 0.258 | 0.690 | 1.337 | 1.746 | 2.120 | 2.583 | 2.921 | 3.252 | 3.686 | 4.015 |
| 17 | 0.257 | 0.689 | 1.333 | 1.740 | 2.110 | 2.567 | 2.898 | 3.222 | 3.646 | 3.965 |
| 18 | 0.257 | 0.688 | 1.330 | 1.734 | 2.101 | 2.552 | 2.878 | 3.197 | 3.610 | 3.922 |
| 19 | 0.257 | 0.688 | 1.328 | 1.729 | 2.093 | 2.539 | 2.861 | 3.174 | 3.579 | 3.883 |
| 20 | 0.257 | 0.687 | 1.325 | 1.725 | 2.086 | 2.528 | 2.845 | 3.153 | 3.552 | 3.850 |
| 21 | 0.257 | 0.686 | 1.323 | 1.721 | 2.080 | 2.518 | 2.831 | 3.135 | 3.527 | 3.819 |
| 22 | 0.256 | 0.686 | 1.321 | 1.717 | 2.074 | 2.508 | 2.819 | 3.119 | 3.505 | 3.792 |
| 23 | 0.256 | 0.685 | 1.319 | 1.714 | 2.069 | 2.500 | 2.807 | 3.104 | 3.485 | 3.767 |
| 24 | 0.256 | 0.685 | 1.318 | 1.711 | 2.064 | 2.492 | 2.797 | 3.091 | 3.467 | 3.745 |
| 25 | 0.256 | 0.684 | 1.316 | 1.708 | 2.060 | 2.485 | 2.787 | 3.078 | 3.450 | 3.725 |
| 26 | 0.256 | 0.684 | 1.315 | 1.706 | 2.056 | 2.479 | 2.779 | 3.067 | 3.435 | 3.707 |
| 27 | 0.256 | 0.684 | 1.314 | 1.703 | 2.052 | 2.473 | 2.771 | 3.057 | 3.421 | 3.690 |
| 28 | 0.256 | 0.683 | 1.313 | 1.701 | 2.048 | 2.467 | 2.763 | 3.047 | 3.408 | 3.674 |
| 29 | 0.256 | 0.683 | 1.311 | 1.699 | 2.045 | 2.462 | 2.756 | 3.038 | 3.396 | 3.659 |
| 30 | 0.256 | 0.683 | 1.310 | 1.697 | 2.042 | 2.457 | 2.750 | 3.030 | 3.385 | 3.646 |
| 40 | 0.255 | 0.681 | 1.303 | 1.684 | 2.021 | 2.423 | 2.704 | 2.971 | 3.307 | 3.551 |
| 60 | 0.254 | 0.679 | 1.296 | 1.671 | 2.000 | 2.390 | 2.660 | 2.915 | 3.232 | 3.460 |
| 120 | 0.254 | 0.677 | 1.289 | 1.658 | 1.980 | 2.358 | 2.617 | 2.860 | 3.160 | 3.373 |
| ∞ | 0.253 | 0.674 | 1.282 | 1.645 | 1.960 | 2.326 | 2.576 | 2.807 | 3.090 | 3.291 |

# 9 Conclusion

The t-value obtained from an unpaired analysis of paired data produces biased results. This is, because the level of correlation between unpaired data is assumed to be zero, and repeated measurements tend to produce a positive correlation. However, negative correlations may sometimes also be observed, if the responders to one treatment are more at risk of being non-responder to a subsequent treatment and vice versa. This chapter examines the use of correlation coefficients for testing paired data, as an alternative to the traditional paired t-test and Wilcoxon test (Chap. 5). The advantage of is that this method unmasks, how the level of correlation between repeated measures affects the overall uncertainty in a study.

Correlations is a phenomenon of major importance in statistical analyses, and must always taken into account.

## 10 Note

More examples of t-tests and linear regression analyses are given in Statistics applied to clinical studies 5th edition, Chaps. 1, 2, 14 and 15, Springer Heidelberg Germany, 2012, from the same authors.

# Chapter 11
# Power Equations

## 1 General Purpose

Power can be described as statistical conclusive force. It can be defined as the chance of finding a difference where there is one. Other chances are the chance of finding no difference where there is one (type II error) and the chance of finding a difference where there is none (type I) error (Chap. 3). A study result is often expressed in the form of the mean result and its standard deviation (SD) or standard error (SE). With the mean result getting larger and the standard error getting smaller, the study will obtain increasing power. This chapter is to show how to compute from a study's mean and standard error its statistical power.

## 2 Schematic Overview of Type of Data File

| Outcome |
|---|
| . |
| . |
| . |
| . |
| . |
| . |
| . |
| . |
| . |
| |
| |
| |
| |



T.J. Cleophas, A.H. Zwinderman, *Clinical Data Analysis on a Pocket Calculator*,
DOI 10.1007/978-3-319-27104-0_11

## 3 Primary Scientific Question

What is the power of a study with its mean study result and its standard error given.

## 4 Power Assessment

Important hypotheses are hypothesis 0 (H0, no difference from a 0 effect), and hypothesis 1 (H1, real difference from a 0 effect). For the purpose of power assessment, we will, particularly, emphasize hypothesis 1. The underneath figure shows graphs of H0 and H1.

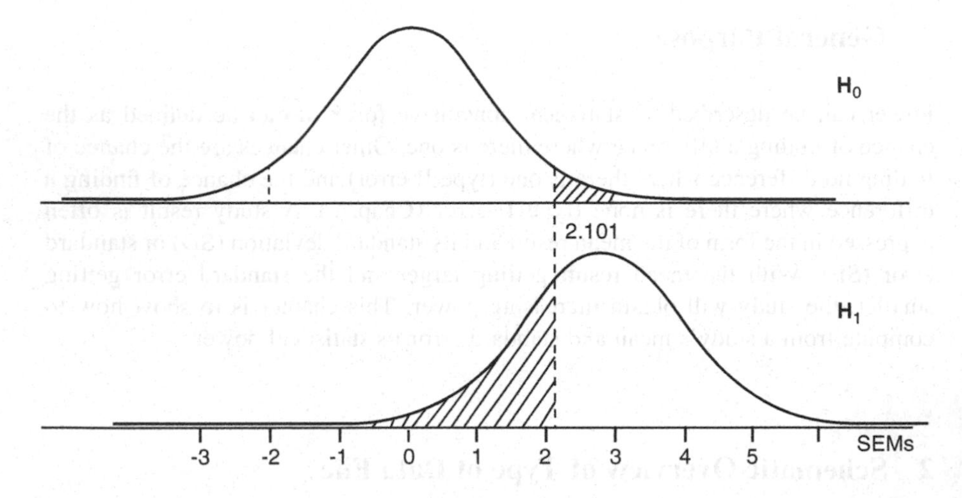


H1 is a graph based on a 2-group trial with a sample size of 20. Mean ± SEMs are on the x-axis (commonly called z-axis here). H0 is the same graph with mean 0. H1 is also the summary of the means many trials similar to ours. H0 is also the summary of the means many trials similar to ours, but with an overall mean effect of 0. If the hypothesis 0 is true, then the mean of our study is part of H0, if the hypothesis 1 is true, then the mean of our study is part of H1. So, the mean of our study may be part of H0, or of H1. We can't prove anything, but we can calculate the chance of either of these possibilities.

A mean result of 2.9 is far distant from 0. Suppose, it belongs to H0. Only 5 % of the H0 trials > 2.1 SEMs distant from 0. The chance, that it belongs to H0 is, thus, < 5 %. Reject this small possibility. Now, suppose the result belongs to H1.

Up to 30 % of the H1 trials are < 2.1 SEMs distant from 0. These 30 % cannot reject null hypothesis of no effect. The trials right from 2.1 SEMs (corresponding with 70 % of the area under the curve (AUC)) can do so.

We can conclude from all of the above considerations: if H0 is true, we will have $< 5\%$ chance to find it; if H1 is true, we will have 70 % chance to find it. And so, we will decide to reject the null hypothesis of no effect at $p < 0.05$, and to do so with a power of 70 %.

# 5 Data Example

A blood pressure study shows a mean decrease in blood pressure of 10.8 mm Hg with a standard error of 3.0 mm Hg. Results from study samples are often given in grams, liters, Euros, mm Hg etc. For the calculation of power we have to first standardize our study result, which means that the mean result has to be divided by its own standard error (SE or SEM):

Mean ± SE =
mean /SE ± SE/SE =
t-value ± 1.

All of the t-values, as found in the t-table, can be looked at as the standardized mean results of all kinds of studies. In our blood pressure study the t-value = 10.8 / 3.0 = 3.6. The unit of the t-value is not mm Hg, but rather SE-units or SEM-units. The question is: what power will the study have, if we assume a type I error (alpha) = 5 % and a sample size of n = 20.

The question is: what will the power of this study be, if we assume a type I error (alpha) of 5 %, and a sample size of n = 20.

A. 90 % < power < 95 %,
B. power > 80 %,
C. power < 75 %,
D. power > 75 %.

n = 20 indicates 20–2 = 18 degrees of freedom in the case of 2 groups of 10 patients.

We will use the following power equation (prob = probability, z = value on the z-line (the x-axis of the t-distribution)

$$\text{Power} = 1 - \text{prob}\left(z < t - t^1\right)$$

| | |
|---|---|
| t | = the t-value of your results, |
| $t^1$ | = the t –value, that matches a p-value of 0.05 = 2.1; |
| t | = 3.6; $t^1 = 2.1$; $t - t^1 = 1.5$; |
| prob $(z < t - t^1)$ | = beta = type II error = between 0.05 and 0.1 |
| 1-beta = power | = between 0.9 and 0.95 = between 90 and 95 %. |

So, there is a very good power here. Explanation of the above calculation is given in the next few lines.

## 6 T-Table

The t-table has a left-end column giving degrees of freedom (≈ sample sizes), and two top rows with p-values (areas under the curve = p-values), one-tail meaning that only one end of the curve, two-tail meaning that both ends are assessed simultaneously. The t-table is, furthermore, full of t-values, that, with ∞ degrees of freedom, are equal to z-values (Chap. 36). The t-values are to be understood as mean results of studies, but not expressed in mmol/l, kilograms, but in so-called SEM-units (Standard error of the mean units), that are obtained by dividing your mean result by its own standard error. With many degrees of freedom (large samples) the curve will be a little bit narrower, and more in agreement with nature. The current chapter shows how the t-table can also be applied for computing statistical power.

With a t-value of 3.6 as shown in the previous section, and 18 degrees of freedom, the term $(t - t^1)$ equals 1.5. This value is between 1.330 and 1.734. Look right up at the upper top row for finding beta (type II error = the chance of finding no difference where there is one). We have two top rows here, one for one-tail testing one for two-tail testing. Power is always tested one-tail. We are between 0.1 and 0.05 (10 and 5 %). This is an adequate estimate of the type II error. The power, thus, equals (100 % – beta) = between 90 and 95 % in our example.

| *df* | One-Tail = .4<br>Two-Tail = .8 | .25<br>.5 | .1<br>.2 | .05<br>.1 | .025<br>.05 | .01<br>.02 | .005<br>.01 | .0025<br>.005 | .001<br>.002 | .0005<br>.001 |
|---|---|---|---|---|---|---|---|---|---|---|
| 1 | 0.325 | 1.000 | 3.078 | 6.314 | 12.706 | 31.821 | 63.657 | 127.32 | 318.31 | 636.62 |
| 2 | 0.289 | 0.816 | 1.886 | 2.920 | 4.303 | 6.965 | 9.925 | 14.089 | 22.327 | 31.598 |
| 3 | 0.277 | 0.765 | 1.638 | 2.353 | 3.182 | 4.541 | 5.841 | 7.453 | 10.214 | 12.924 |
| 4 | 0.271 | 0.741 | 1.533 | 2.132 | 2.776 | 3.747 | 4.604 | 5.598 | 7.173 | 8.610 |
| 5 | 0.267 | 0.727 | 1.476 | 2.015 | 2.571 | 3.365 | 4.032 | 4.773 | 5.893 | 6.869 |
| 6 | 0.265 | 0.718 | 1.440 | 1.943 | 2.447 | 3.143 | 3.707 | 4.317 | 5.208 | 5.959 |
| 7 | 0.263 | 0.711 | 1.415 | 1.895 | 2.365 | 2.998 | 3.499 | 4.029 | 4.785 | 5.408 |
| 8 | 0.262 | 0.706 | 1.397 | 1.860 | 2.306 | 2.896 | 3.355 | 3.833 | 4.501 | 5.041 |
| 9 | 0.261 | 0.703 | 1.383 | 1.833 | 2.262 | 2.821 | 3.250 | 3.690 | 4.297 | 4.781 |
| 10 | 0.260 | 0.700 | 1.372 | 1.812 | 2.228 | 2.764 | 3.169 | 3.581 | 4.144 | 4.587 |
| 11 | 0.260 | 0.697 | 1.363 | 1.796 | 2.201 | 2.718 | 3.106 | 3.497 | 4.025 | 4.437 |
| 12 | 0.259 | 0.695 | 1.356 | 1.782 | 2.179 | 2.681 | 3.055 | 3.428 | 3.930 | 4.318 |
| 13 | 0.259 | 0.694 | 1.350 | 1.771 | 2.160 | 2.650 | 3.012 | 3.372 | 3.852 | 4.221 |
| 14 | 0.258 | 0.692 | 1.345 | 1.761 | 2.145 | 2.624 | 2.977 | 3.326 | 3.787 | 4.140 |
| 15 | 0.258 | 0.691 | 1.341 | 1.753 | 2.131 | 2.602 | 2.947 | 3.286 | 3.733 | 4.073 |
| 16 | 0.258 | 0.690 | 1.337 | 1.746 | 2.120 | 2.583 | 2.921 | 3.252 | 3.686 | 4.015 |
| 17 | 0.257 | 0.689 | 1.333 | 1.740 | 2.110 | 2.567 | 2.898 | 3.222 | 3.646 | 3.965 |
| 18 | 0.257 | 0.688 | 1.330 | 1.734 | 2.101 | 2.552 | 2.878 | 3.197 | 3.610 | 3.922 |
| 19 | 0.257 | 0.688 | 1.328 | 1.729 | 2.093 | 2.539 | 2.861 | 3.174 | 3.579 | 3.883 |
| 20 | 0.257 | 0.687 | 1.325 | 1.725 | 2.086 | 2.528 | 2.845 | 3.153 | 3.552 | 3.850 |
| 21 | 0.257 | 0.686 | 1.323 | 1.721 | 2.080 | 2.518 | 2.831 | 3.135 | 3.527 | 3.819 |
| 22 | 0.256 | 0.686 | 1.321 | 1.717 | 2.074 | 2.508 | 2.819 | 3.119 | 3.505 | 3.792 |
| 23 | 0.256 | 0.685 | 1.319 | 1.714 | 2.069 | 2.500 | 2.807 | 3.104 | 3.485 | 3.767 |
| 24 | 0.256 | 0.685 | 1.318 | 1.711 | 2.064 | 2.492 | 2.797 | 3.091 | 3.467 | 3.745 |
| 25 | 0.256 | 0.684 | 1.316 | 1.708 | 2.060 | 2.485 | 2.787 | 3.078 | 3.450 | 3.725 |
| 26 | 0.256 | 0.684 | 1.315 | 1.706 | 2.056 | 2.479 | 2.779 | 3.067 | 3.435 | 3.707 |
| 27 | 0.256 | 0.684 | 1.314 | 1.703 | 2.052 | 2.473 | 2.771 | 3.057 | 3.421 | 3.690 |
| 28 | 0.256 | 0.683 | 1.313 | 1.701 | 2.048 | 2.467 | 2.763 | 3.047 | 3.408 | 3.674 |
| 29 | 0.256 | 0.683 | 1.311 | 1.699 | 2.045 | 2.462 | 2.756 | 3.038 | 3.396 | 3.659 |
| 30 | 0.256 | 0.683 | 1.310 | 1.697 | 2.042 | 2.457 | 2.750 | 3.030 | 3.385 | 3.646 |
| 40 | 0.255 | 0.681 | 1.303 | 1.684 | 2.021 | 2.423 | 2.704 | 2.971 | 3.307 | 3.551 |
| 60 | 0.254 | 0.679 | 1.296 | 1.671 | 2.000 | 2.390 | 2.660 | 2.915 | 3.232 | 3.460 |
| 120 | 0.254 | 0.677 | 1.289 | 1.658 | 1.980 | 2.358 | 2.617 | 2.860 | 3.160 | 3.373 |
| ∞ | 0.253 | 0.674 | 1.282 | 1.645 | 1.960 | 2.326 | 2.576 | 2.807 | 3.090 | 3.291 |

# 7 Conclusion

Power can be defined as the chance of finding an effect (or a difference from zero), where there is one. It is equal to 1 minus the type II error ($=1-\beta$). A study result is often expressed in the form of the mean result and its standard deviation (SD) or standard error (SE or SEM). With the mean result getting larger and the standard error getting smaller, the study will obtain increasing power. This chapter shows, how to compute a study's statistical power from its mean and standard error.

# 8 Note

More background, theoretical and mathematical information of power assessments is given in Statistics applied to clinical studies 5th edition, Chap. 6, Springer Heidelberg Germany, 2012, from the same authors.

# Chapter 12
# Sample Size Calculations

## 1 General Purpose

When writing a study protocol, just pulling the sample size out of a hat gives rise to (1) ethical, (2) scientific, and (3) financial problems, because

1. too many patients may be given a potentially inferior treatment,
2. negative studies require repetition of the research,
3. costs are involved in too large or too small studies.

An essential part of preparing clinical studies is the question, how many subjects need to be studied in order to answer the studies' objectives. This chapter provides equations that can be used for the purpose.

## 2 Schematic Overview of Type of Data File

| Outcome |
| --- |
| . |
| . |
| . |
| . |
| . |
| . |
| . |
| . |
| . |
| |
| |
| |
| |



T.J. Cleophas, A.H. Zwinderman, *Clinical Data Analysis on a Pocket Calculator*,
DOI 10.1007/978-3-319-27104-0_12

## 3 Primary Scientific Question

What sample size do we need in order to produce a study with a statistically significant result?

## 4 Data Example, Continuous Data, Power 50 %

An essential part of clinical studies is the question, how many subject need to be studied in order to answer the studies' objectives. As an example, we will use an intended study that has an expected mean effect of 5, and a standard deviation (SD) of 15.

What required sample size do we need in order to obtain a significant result, or, in other words, to obtain a p-value of at least 0.05.

A. 16,
B. 36,
C. 64,
D. 100.

A suitable equation to assess this question can be constructed as follows. With a study's t-value of 2.0 SEM-units (SEM = standard error of the mean), a significant p-value of 0.05 will be obtained. This should not be difficult for you to understand, when you think of the 95 % confidence interval of the mean of a study being between – 2 and + 2 SEM-units (Chap. 13).

We assume

t-value = 2 SEMs
= (mean study result) / (standard error)
= (mean study result) / (standard deviation /√n)

(n = study's sample size)

From the above equation it can be derived that

√n = 2 × standard deviation (SD) / (mean study result)
n = required sample size
= 4 × (SD/(mean study result))$^2$
= 4 × (15 / 5) $^2$ = 36

Answer B is correct.

You are testing here whether a result of 5 is significantly different from a result of 0. Often two groups of data are compared and the standard deviations of the two groups have to be pooled (see Chap. 7). As stated above, with a t-value of 2.0 SEMs a significant result of p = 0.05 will be obtained. However, the power of this study is only 50 %, indicating, that you will have a power of only 50 % (= the chance of an insignificant result the next time you perform a similar study).

## 5 Data Example, Continuous Data, Power 80 %

What is the required sample size of a study with an expected mean result of 5, and SD of 15, and, that should have a p-value of at least 0.05 and a power of at least 80 % (power index = $(z_{\alpha} + z_{\beta})^2 = 7.8$).

A. 140,
B. 70,
C. 280,
D. 420.

An adequate equation is the following.

Required sample size $= \text{power index} \times (\text{SD/mean})^2$
$= 7.8 \times (15 / 5)^2 = 70$

If you wish to have a power in your study of 80 % instead of 50 %, you will need a larger sample size. With a power of only 50 % your required sample size was only 36.

## 6 Data Example, Continuous Data, Power 80 %, Two Groups

What is the required sample size of a study with two groups and a mean difference of 5 and SDs of 15 per Group, and that will have a p-value of at least 0.05 and a power of at least 80 % (Power index = $(z_{\alpha} + z_{\beta})^2 = 7.8$).

A. 140,
B. 70,
C. 280,
D. 420.

The suitable equation is given underneath.

Required sample size $= \text{power index} \times (\text{pooled SD})^2 / (\text{mean difference})^2$.
$(\text{pooled SD})^2 = SD_1^2 + SD_2^2$.
Required sample size $= 7.8 \times (15^2 + 15^2) / 5^2 = 140$.

The required sample size is 140 patients per group. And so, with two groups you will need considerably larger samples than you will with 1 group.

## 7 Conclusion

When writing a study protocol, just pulling the sample size out of a hat gives rise to (1) ethical, (2) scientific, and (3) financial problems. An essential part of preparing clinical studies is the question, how many subject need to be studied in order to answer the studies' objectives. Equations are provided, that can be used for the purpose.

## 8 Note

More background, theoretical and mathematical information of sample size requirements is given in Statistics applied to clinical studies 5th edition, Chap. 6, Springer Heidelberg Germany, 2012, from the same authors.

# Chapter 13
# Confidence Intervals

## 1 General Purpose

The 95% confidence interval of a study represents an interval covering 95% of the means of many studies similar to that of our study. It tells you something about what you can expect from future data: if you repeat the study, you will be 95% sure that your mean outcome will be within the 95% confidence interval. The chapter shows how it can be calculated.

## 2 Schematic Overview of Type of Data File

| Outcome |
|---|
| . |
| . |
| . |
| . |
| . |
| . |
| . |
| . |
| . |
| |
| |
| |
| |
| |



T.J. Cleophas, A.H. Zwinderman, *Clinical Data Analysis on a Pocket Calculator*,
DOI 10.1007/978-3-319-27104-0_13

## 3 Primary Scientific Question

Can the 95% confidence intervals be used as an alternative to statistical significance testing? What are the advantages?

## 4 Data Example, Continuous Outcome Data

The 95% confidence interval of a study represents an interval covering 95% of the means of many studies similar to that of our study. It tells you something about what you can expect from future data: if you repeat the study, you will be 95% sure that the outcome will be within the 95% confidence interval. The 95% confidence of a study is found by the equation

95% confidence interval = mean $\pm$ 2 x standard error (SE)

The SE is equal to the standard deviation (SD) / $\sqrt{n}$, where n = the sample size of your study. The SD can be calculated from the procedure reviewed in the Chap. 2.

With an SD of 1.407885953 and a sample size of n = 8,

your SE = 1.407885953 / $\sqrt{8}$
= 0.4977,

with a mean value of your study of 53.375,
with a 95% confidence interval = 53.375 $\pm$ 2 $\times$ 0.4977
= between 52.3796 and 54.3704.

The mean study results are often reported together with 95% confidence intervals. They are also the basis for equivalence studies and noninferiority studies, which will be reviewed in the Chaps. 14 and 15. Also for study results expressed in the form of numbers of events, proportion of deaths, odds ratios of events, etc., 95 % confidence intervals can be readily calculated.

We should add that the equation

95% confidence interval = mean $\pm$ 2 $\times$ standard error (SE),

is a pretty rough approximation, and that a more precise estimate would be the equation

95% confidence interval = mean $\pm$ $t^1$ $\times$ standard error (SE),

where $t^1$ = the critical t-value corresponding to a two-sided p-value of 0.05.

# 5 T-Table and 95% Confidence Intervals

The t-table has a left-end column giving degrees of freedom (≈ sample sizes), and two top rows with p-values (areas under the curve = p-values), one-tail meaning that only one end of the curve, two-tail meaning that both ends are assessed simultaneously. The t-table is, furthermore, full of t-values, that, with ∞ degrees of freedom, are equal to z-values (Chap. 36). The t-values are to be understood as mean results of studies, but not expressed in mmol/l, kilograms, but in so-called SEM-units (Standard error of the mean units), that are obtained by dividing your mean result by its own standard error. With many degrees of freedom (large samples) the curve will be a little bit narrower, and more in agreement with nature.

| | One-Tail = .4 | .25 | .1 | .05 | .025 | .01 | .005 | .0025 | .001 | .0005 |
|---|---|---|---|---|---|---|---|---|---|---|
| *df* | Two-Tail = .8 | .5 | .2 | .1 | .05 | .02 | .01 | .005 | .002 | .001 |
| **1** | 0.325 | 1.000 | 3.078 | 6.314 | 12.706 | 31.821 | 63.657 | 127.32 | 318.31 | 636.62 |
| **2** | 0.289 | 0.816 | 1.886 | 2.920 | 4.303 | 6.965 | 9.925 | 14.089 | 22.327 | 31.598 |
| **3** | 0.277 | 0.765 | 1.638 | 2.353 | 3.182 | 4.541 | 5.841 | 7.453 | 10.214 | 12.924 |
| **4** | 0.271 | 0.741 | 1.533 | 2.132 | 2.776 | 3.747 | 4.604 | 5.598 | 7.173 | 8.610 |
| **5** | 0.267 | 0.727 | 1.476 | 2.015 | 2.571 | 3.365 | 4.032 | 4.773 | 5.893 | 6.869 |
| **6** | 0.265 | 0.718 | 1.440 | 1.943 | 2.447 | 3.143 | 3.707 | 4.317 | 5.208 | 5.959 |
| **7** | 0.263 | 0.711 | 1.415 | 1.895 | 2.365 | 2.998 | 3.499 | 4.029 | 4.785 | 5.408 |
| **8** | 0.262 | 0.706 | 1.397 | 1.860 | 2.306 | 2.896 | 3.355 | 3.833 | 4.501 | 5.041 |
| **9** | 0.261 | 0.703 | 1.383 | 1.833 | 2.262 | 2.821 | 3.250 | 3.690 | 4.297 | 4.781 |
| **10** | 0.260 | 0.700 | 1.372 | 1.812 | 2.228 | 2.764 | 3.169 | 3.581 | 4.144 | 4.587 |
| **11** | 0.260 | 0.697 | 1.363 | 1.796 | 2.201 | 2.718 | 3.106 | 3.497 | 4.025 | 4.437 |
| **12** | 0.259 | 0.695 | 1.356 | 1.782 | 2.179 | 2.681 | 3.055 | 3.428 | 3.930 | 4.318 |
| **13** | 0.259 | 0.694 | 1.350 | 1.771 | 2.160 | 2.650 | 3.012 | 3.372 | 3.852 | 4.221 |
| **14** | 0.258 | 0.692 | 1.345 | 1.761 | 2.145 | 2.624 | 2.977 | 3.326 | 3.787 | 4.140 |
| **15** | 0.258 | 0.691 | 1.341 | 1.753 | 2.131 | 2.602 | 2.947 | 3.286 | 3.733 | 4.073 |
| **16** | 0.258 | 0.690 | 1.337 | 1.746 | 2.120 | 2.583 | 2.921 | 3.252 | 3.686 | 4.015 |
| **17** | 0.257 | 0.689 | 1.333 | 1.740 | 2.110 | 2.567 | 2.898 | 3.222 | 3.646 | 3.965 |
| **18** | 0.257 | 0.688 | 1.330 | 1.734 | 2.101 | 2.552 | 2.878 | 3.197 | 3.610 | 3.922 |
| **19** | 0.257 | 0.688 | 1.328 | 1.729 | 2.093 | 2.539 | 2.861 | 3.174 | 3.579 | 3.883 |
| **20** | 0.257 | 0.687 | 1.325 | 1.725 | 2.086 | 2.528 | 2.845 | 3.153 | 3.552 | 3.850 |
| **21** | 0.257 | 0.686 | 1.323 | 1.721 | 2.080 | 2.518 | 2.831 | 3.135 | 3.527 | 3.819 |
| **22** | 0.256 | 0.686 | 1.321 | 1.717 | 2.074 | 2.508 | 2.819 | 3.119 | 3.505 | 3.792 |
| **23** | 0.256 | 0.685 | 1.319 | 1.714 | 2.069 | 2.500 | 2.807 | 3.104 | 3.485 | 3.767 |
| **24** | 0.256 | 0.685 | 1.318 | 1.711 | 2.064 | 2.492 | 2.797 | 3.091 | 3.467 | 3.745 |
| **25** | 0.256 | 0.684 | 1.316 | 1.708 | 2.060 | 2.485 | 2.787 | 3.078 | 3.450 | 3.725 |
| **26** | 0.256 | 0.684 | 1.315 | 1.706 | 2.056 | 2.479 | 2.779 | 3.067 | 3.435 | 3.707 |
| **27** | 0.256 | 0.684 | 1.314 | 1.703 | 2.052 | 2.473 | 2.771 | 3.057 | 3.421 | 3.690 |
| **28** | 0.256 | 0.683 | 1.313 | 1.701 | 2.048 | 2.467 | 2.763 | 3.047 | 3.408 | 3.674 |
| **29** | 0.256 | 0.683 | 1.311 | 1.699 | 2.045 | 2.462 | 2.756 | 3.038 | 3.396 | 3.659 |
| **30** | 0.256 | 0.683 | 1.310 | 1.697 | 2.042 | 2.457 | 2.750 | 3.030 | 3.385 | 3.646 |
| **40** | 0.255 | 0.681 | 1.303 | 1.684 | 2.021 | 2.423 | 2.704 | 2.971 | 3.307 | 3.551 |
| **60** | 0.254 | 0.679 | 1.296 | 1.671 | 2.000 | 2.390 | 2.660 | 2.915 | 3.232 | 3.460 |
| **120** | 0.254 | 0.677 | 1.289 | 1.658 | 1.980 | 2.358 | 2.617 | 2.860 | 3.160 | 3.373 |
| **∞** | 0.253 | 0.674 | 1.282 | 1.645 | 1.960 | 2.326 | 2.576 | 2.807 | 3.090 | 3.291 |

In the fifth column of t-values of the above t-table all of the $t^1$-values are given. For example, with a sample of 120 the $t^1$-value equals 1.980, with a sample size of close to 8 the $t^1$-value rise to 2.306.

# 6 Data Example, Binary Outcome Data

What is the standard error (SE) of a study with events in 10% of the patients, and a sample size of 100 (n). Ten % events means a proportion of events of 0.1. The standard deviation (SD) of this proportion is defined by the equation

$\sqrt{}$ [proportion × (1 − proportion)] $= \sqrt{}(0.1 \times 0.9) = \sqrt{}0.09 = 0.3$,

the standard error = standard deviation/$\sqrt{}$n,
= 0.3/10 = 0.03,

the 95 % confidence interval is given by
proportion given ± 1.960 × 0.03 = 0.1 ± 1.960 × 0.03,
= 0.1 ± 0.06,
= between 0.04 and 0.16.

# 7 Conclusion

The 95% confidence interval of a study represents an interval covering 95% of the means of many studies similar to that of our study. It tells you something about what you can expect from future data: if you repeat the study, you will be 95% sure that your mean outcome will be within the 95% confidence interval. The 95% confidence interval can be used as an alternative to statistical significance testing. The advantages are that the intervals picture expected mean results of future data, and that they can be applied for studying therapeutic equivalence and noninferiority (Chaps. 14 and 15).

# 8 Note

More background, theoretical and mathematical information of confidence intervals are given in the Chaps. 14 and 15 of this volume.

# Chapter 14
# Equivalence Testing Instead of Null-Hypothesis Testing

## 1 General Purpose

A negative study is not equal to an equivalent study. The former can not reject the null-hypothesis of no effect, while the latter assesses whether its 95 % confidence interval is between prior boundaries, defining an area of undisputed clinical relevance. Equivalence testing is important, if you expect a new treatment to be equally efficaceous as the standard treatment. This new treatment may still be better suitable for practice, if it has fewer adverse effects or other ancillary advantages. For the purpose of equivalence testing we need to set boundaries of equivalence prior to the study. After the study we check whether the 95 % confidence interval of the study is

1. entirely within the boundaries (equivalence is demonstrated),
2. partly within (equivalence is unsure),
3. entirely without (equivalence is ruled out).

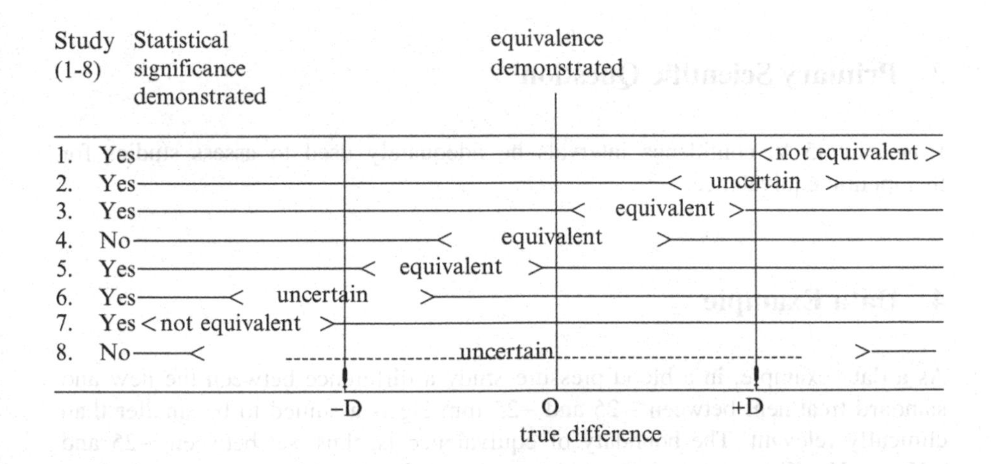




T.J. Cleophas, A.H. Zwinderman, *Clinical Data Analysis on a Pocket Calculator*,
DOI 10.1007/978-3-319-27104-0_14

The above figure gives between brackets both the 95 % confidence intervals of 8 studies, and the defined boundaries of equivalence (−D and +D boundary):

The studies 1–8:
1 and 7 are completely without the boundaries (no equivalence),
2, 6 and 8 are partly without the boundaries (equivalence is unsure),
3–5 are entirely within the boundaries (equivalence is demonstrated).

Particularly, the studies 3 and 5 are remarkable, because they show simultaneously the presence of clinical equivalence and of a statistically significant difference from a zero effect.

## 2 Schematic Overview of Type of Data File

| Outcome |
|---|
| . |
| . |
| . |
| . |
| . |
| . |
| . |
| . |
| . |
| |
| |
| |

## 3 Primary Scientific Question

How can 95 % confidence intervals be adequately used to assess studies for therapeutic equivalence.

## 4 Data Example

As a data example, in a blood pressure study a difference between the new and standard treatment between −25 and +25 mm Hg is assumed to be smaller than clinically relevant. The boundary of equivalence is, thus, set between −25 and +25 mm Hg. This boundary should be a priori defined in the study protocol.

Then, the study is carried out, and the new and the standard treatment produce a mean reduction in blood pressure of 9.9 and 8.4 mm Hg (parallel-group study of 20 patients) with standard errors of 7.0 and 6.9 mm Hg.

The mean difference = 9.9 minus 8.4 mm Hg
=1.5 mm Hg

The standard errors of the mean differences are 7.0 and 6.9 mm Hg

The pooled standard error = $\sqrt{}$ $(7.0^2 + 6.9^2)$ mm Hg
= $\sqrt{}$ 96.61 mm Hg
= 9.83 mm Hg

The 95 % confidence interval of this study = $1.5 \pm 2.0 \times 9.83$ mm Hg
= between −18.16 and +21.16 mm Hg

This result is entirely within the a priori defined boundary of equivalence, which means that equivalence is demonstrated in this study.

## 5 Conclusion

A negative study is not equal to an equivalent study. The former assesses the null-hypothesis of no effect, while the latter assesses, whether its 95 % confidence interval is between a priori defined boundaries, defining an area of undisputed clinical relevance. Equivalence testing is important, if you expect a new treatment to be equally efficaceous as the standard treatment. This new treatment may still be better suitable for practice, if it has fewer adverse effects or other ancillary advantages. For the purpose of equivalence testing we need to set boundaries of equivalence prior to the study. The boundaries of equivalence must be in the protocol, and equivalence after the study has been completed is impossible. In an equivalence study, after the study has been completed, you should check, whether the 95 % confidence interval of the study is entirely within the a priori defined boundaries of equivalence. The boundaries have been defined on clinical, not statistical grounds.

In the current chapter, a study with continuous outcome data is used as an example. When studying binary outcome data, the result is often expressed as the proportion responders, e.g., 0.4 or 40 % responders. The calculation of the standard error with binary outcomes is explained in the Chap. 37. Briefly, with a proportion of 0.4 responders and a study sample size of 100, the standard error equals

SE = $\sqrt{[(0.4x0.6)/100]} = 0.049$

The confidence interval of this study = $0.4 \pm 1.960 \times 0.049$
= between 0.304 and 0.496

If the prior boundaries of equivalence were defined as being a proportion of responders between 0.25 and 0.50, then this study demonstrates the presence of equivalence.

## 6 Note

More background, theoretical and mathematical information of equivalence testing is given in Statistics applied to clinical studies 5th edition, Chap. 5, Springer Heidelberg Germany, from the same authors.

# Chapter 15
# Noninferiority Testing Instead of Null-Hypothesis Testing

## 1 General Purpose

Just like equivalence studies (Chap. 14), noninferiority studies are very popular in modern clinical research, with many treatments at hand, and with new compounds being mostly only slightly different from the old ones. Unlike equivalence studies, noninferiority studies have, instead of two boundaries with an interval of equivalence in between, a single boundary. Noninferiority studies have been criticized for their wide margin of inferiority, making it, virtually, impossible to reject noninferiority.

The underneath graph shows the possible results in the form of confidence intervals of three noninferiority trials that have the same boundary or margin of noninferiority. This chapter is to provide a procedure to adequately analyze noninferiority trials.

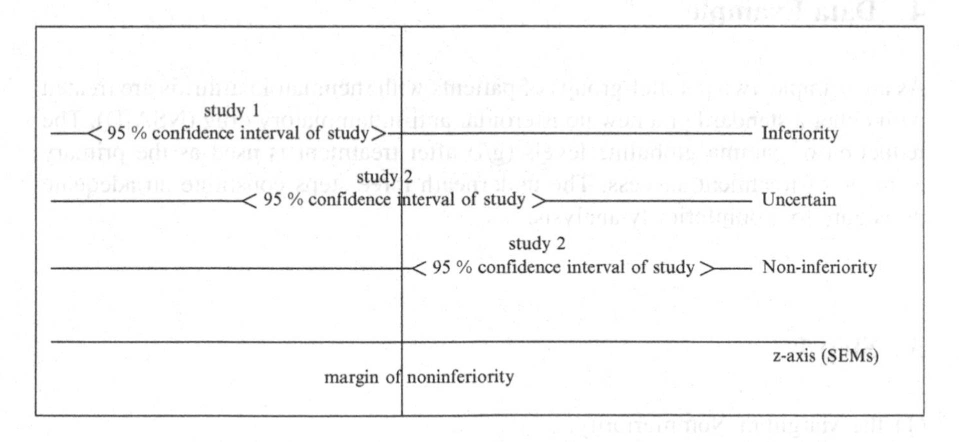




T.J. Cleophas, A.H. Zwinderman, *Clinical Data Analysis on a Pocket Calculator*,
DOI 10.1007/978-3-319-27104-0_15

## 2 Schematic Overview of Type of Data File

| Outcome |
|---|
| . |
| . |
| . |
| . |
| . |
| . |
| . |
| . |
| . |

## 3 Primary Scientific Question

Is a novel treatment, if not significantly better than a standard treatment, significantly noninferior as compared to the standard treatment.

## 4 Data Example

As an example, two parallel-groups of patients with rheumatoid arthritis are treated with either a standard or a new nonsteroidal anti-inflammatory drug (NSAID). The reduction of gamma globuline levels (g/l) after treatment is used as the primary estimate of treatment success. The underneath three steps constitute an adequate procedure for noninferiority analysis.

## 5 Step 1

(1) the Margin of Noninferiority,
(2) the Required Sample Size,
(3) the Expected P-Value and Power

(1)

The left boundaries of the 95 % confidence intervals of previously published studies of the standard NSAID versus various alternative NSAIDS were never lower than –8 g/l. And, so, the margin was set at −8 g/l.

(2)

Based on a pilot-study with the novel compound the expected mean difference was 0 g/l with an expected standard deviation of 32 g/l. This would mean a required sample size of

$n = \text{power index} \times (\text{SD}/(\text{margin} - \text{mean}))^2$
$n = 7.8 \times (32/(-8-0))^2 = 125$ patients per group.

A power index of 7.8 takes care that noninferiority is demonstrated with a power of about 80% in this study (see also Chap. 11).

(3)

The mean difference between the new and standard NSAID was calculated to be 3.0 g/l with a standard error (SE) of 4.6 g/l. This means that the t-value of the study equaled $t = (\text{margin} - \text{mean})/\text{SE} = (-8\ -3)/4.6 = -2.39$ SE-units or SEM-units.

This t-value corresponds with a p-value of <0.05 (see Chap. 4). Non-inferiority is, thus, demonstrated at $p < 0.05$.

# 6 Step 2

Testing the Significance of Difference Between the New and the Standard Treatment

The mean difference between the new and standard treatment equaled 3.0 g/l with an SE of 4.6 g/l. The 95% confidence of this result is $3.0 \pm 2*4.6$, and is between -6.2 and 12.2 g/l (* = sign of multiplication). This interval does cross the zero value on the z-axis, which means no significant difference from zero ($p > 0.05$).

# 7 Step 3

Testing the Significance of Difference Between the New Treatment And a Placebo

A similarly sized published trial of the standard treatment versus placebo produced a t-value of 2.83, and, thus a p-value of 0.0047. The t-value of the current trial equals $3.0/4.6 = 0.65$ SE-units. The add-up sum $2.83 + 0.65 = 3.48$ is an adequate estimate of the t-value for the comparison of the new treatment versus placebo. A t-value of 3.48 corresponds with a p-value of $< 0.001$ (see Chap. 4). This would mean that the new treatment is significantly better than placebo at $p < 0.001$.

## 8 Conclusion

We can now conclude that

(1) noninferiority is demonstrated at $p < 0.05$,
(2) a significant difference between the new and standard treatment is rejected at $p > 0.05$,
(3) the new treatment is significantly better than placebo at $p < 0.001$.

Non-inferiority has, thus, been unequivocally demonstrated in this study.

Also studies with binary outcomes have 95% confidence intervals (as shown in the Chap. 13), and can, thus, be tested for noninferiority.

## 9 Note

More background, theoretical and mathematical information of noninferiority testing is given in Statistics applied to clinical studies 5th edition, Chap. 63, Springer Heidelberg Germany, 2012, from the same authors.

# Chapter 16
# Superiority Testing Instead of Null-Hypothesis Testing

## 1 General Purpose

For this chapter some knowledge of power equations is required. This is given in the Chaps. 11 and 50. Superiority testing of a study means testing whether the study meets its a priori defined expected power. Many therapeutic studies may be able to reject their null-hypotheses, and, are, thus, statistically significant, but they do not meet their expected power. Although p-values are widely reported, power is rarely given in the report. This may be a problem in practice, since lack of power indicates that the treatments are less efficaceous than expected. Superiority testing assesses whether the eventual power of a study is in agreement with the power as stated in the protocol of the study. This chapter shows how superiority can be assessed.

## 2 Schematic Overview of Type of Data File

| Outcome | predictor |
|---|---|
| . | . |
| . | . |
| . | . |
| . | . |
| . | . |
| . | . |
| . | . |
| . | . |
| . | . |
| | |
| | |
| | |



T.J. Cleophas, A.H. Zwinderman, *Clinical Data Analysis on a Pocket Calculator*,
DOI 10.1007/978-3-319-27104-0_16

## 3 Primary Scientific Question

Is the expected power level as assessed prior to a study in agreement with the power level obtained.

## 4 Data Example

The expected power of a study of a 10 patient crossover study is 90 %. The results of the study are given underneath:

observation 1:
6.0, 7.1, 8.1, 7.5, 6.4, 7.9, 6.8, 6.6, 7.3, 5.6

observation 2:
5.1, 8.0, 3.8, 4.4, 5.2, 5.4, 4.3, 6.0, 3.7, 6.2

Individual differences
0.9, -0.9, 4.3, 3.1, 1.2, 2.5, 2.5, 0.6, 3.8, -0.6

Is there a significant difference between the observation 1 and 2, and which level of significance is correct?

Mean difference = 1.59
SD of mean difference = 1.789
SE = SD/√10 = 0.566
t = 1.59/0.566 = 2.809
10 − 1 = 9 degrees of freedom (10 patients and 1 group of patients).

Look at the underneath t-table to find the p-value, and assess the presence of superiority.

# 5 T-Table

| df | One-Tail = .4<br>Two-Tail = .8 | .25<br>.5 | .1<br>.2 | .05<br>.1 | .025<br>.05 | .01<br>.02 | .005<br>.01 | .0025<br>.005 | .001<br>.002 | .0005<br>.001 |
|---|---|---|---|---|---|---|---|---|---|---|
| 1 | 0.325 | 1.000 | 3.078 | 6.314 | 12.706 | 31.821 | 63.657 | 127.32 | 318.31 | 636.62 |
| 2 | 0.289 | 0.816 | 1.886 | 2.920 | 4.303 | 6.965 | 9.925 | 14.089 | 22.327 | 31.598 |
| 3 | 0.277 | 0.765 | 1.638 | 2.353 | 3.182 | 4.541 | 5.841 | 7.453 | 10.214 | 12.924 |
| 4 | 0.271 | 0.741 | 1.533 | 2.132 | 2.776 | 3.747 | 4.604 | 5.598 | 7.173 | 8.610 |
| 5 | 0.267 | 0.727 | 1.476 | 2.015 | 2.571 | 3.365 | 4.032 | 4.773 | 5.893 | 6.869 |
| 6 | 0.265 | 0.718 | 1.440 | 1.943 | 2.447 | 3.143 | 3.707 | 4.317 | 5.208 | 5.959 |
| 7 | 0.263 | 0.711 | 1.415 | 1.895 | 2.365 | 2.998 | 3.499 | 4.029 | 4.785 | 5.408 |
| 8 | 0.262 | 0.706 | 1.397 | 1.860 | 2.306 | 2.896 | 3.355 | 3.833 | 4.501 | 5.041 |
| 9 | 0.261 | 0.703 | 1.383 | 1.833 | 2.262 | 2.821 | 3.250 | 3.690 | 4.297 | 4.781 |
| 10 | 0.260 | 0.700 | 1.372 | 1.812 | 2.228 | 2.764 | 3.169 | 3.581 | 4.144 | 4.587 |
| 11 | 0.260 | 0.697 | 1.363 | 1.796 | 2.201 | 2.718 | 3.106 | 3.497 | 4.025 | 4.437 |
| 12 | 0.259 | 0.695 | 1.356 | 1.782 | 2.179 | 2.681 | 3.055 | 3.428 | 3.930 | 4.318 |
| 13 | 0.259 | 0.694 | 1.350 | 1.771 | 2.160 | 2.650 | 3.012 | 3.372 | 3.852 | 4.221 |
| 14 | 0.258 | 0.692 | 1.345 | 1.761 | 2.145 | 2.624 | 2.977 | 3.326 | 3.787 | 4.140 |
| 15 | 0.258 | 0.691 | 1.341 | 1.753 | 2.131 | 2.602 | 2.947 | 3.286 | 3.733 | 4.073 |
| 16 | 0.258 | 0.690 | 1.337 | 1.746 | 2.120 | 2.583 | 2.921 | 3.252 | 3.686 | 4.015 |
| 17 | 0.257 | 0.689 | 1.333 | 1.740 | 2.110 | 2.567 | 2.898 | 3.222 | 3.646 | 3.965 |
| 18 | 0.257 | 0.688 | 1.330 | 1.734 | 2.101 | 2.552 | 2.878 | 3.197 | 3.610 | 3.922 |
| 19 | 0.257 | 0.688 | 1.328 | 1.729 | 2.093 | 2.539 | 2.861 | 3.174 | 3.579 | 3.883 |
| 20 | 0.257 | 0.687 | 1.325 | 1.725 | 2.086 | 2.528 | 2.845 | 3.153 | 3.552 | 3.850 |
| 21 | 0.257 | 0.686 | 1.323 | 1.721 | 2.080 | 2.518 | 2.831 | 3.135 | 3.527 | 3.819 |
| 22 | 0.256 | 0.686 | 1.321 | 1.717 | 2.074 | 2.508 | 2.819 | 3.119 | 3.505 | 3.792 |
| 23 | 0.256 | 0.685 | 1.319 | 1.714 | 2.069 | 2.500 | 2.807 | 3.104 | 3.485 | 3.767 |
| 24 | 0.256 | 0.685 | 1.318 | 1.711 | 2.064 | 2.492 | 2.797 | 3.091 | 3.467 | 3.745 |
| 25 | 0.256 | 0.684 | 1.316 | 1.708 | 2.060 | 2.485 | 2.787 | 3.078 | 3.450 | 3.725 |
| 26 | 0.256 | 0.684 | 1.315 | 1.706 | 2.056 | 2.479 | 2.779 | 3.067 | 3.435 | 3.707 |
| 27 | 0.256 | 0.684 | 1.314 | 1.703 | 2.052 | 2.473 | 2.771 | 3.057 | 3.421 | 3.690 |
| 28 | 0.256 | 0.683 | 1.313 | 1.701 | 2.048 | 2.467 | 2.763 | 3.047 | 3.408 | 3.674 |
| 29 | 0.256 | 0.683 | 1.311 | 1.699 | 2.045 | 2.462 | 2.756 | 3.038 | 3.396 | 3.659 |
| 30 | 0.256 | 0.683 | 1.310 | 1.697 | 2.042 | 2.457 | 2.750 | 3.030 | 3.385 | 3.646 |
| 40 | 0.255 | 0.681 | 1.303 | 1.684 | 2.021 | 2.423 | 2.704 | 2.971 | 3.307 | 3.551 |
| 60 | 0.254 | 0.679 | 1.296 | 1.671 | 2.000 | 2.390 | 2.660 | 2.915 | 3.232 | 3.460 |
| 120 | 0.254 | 0.677 | 1.289 | 1.658 | 1.980 | 2.358 | 2.617 | 2.860 | 3.160 | 3.373 |
| ∞ | 0.253 | 0.674 | 1.282 | 1.645 | 1.960 | 2.326 | 2.576 | 2.807 | 3.090 | 3.291 |

The t-table has a left-end column giving degrees of freedom (≈ sample sizes), and two top rows with p-values (areas under the curve = p – values), one-tail meaning that only one end of the curve, two-tail meaning that both ends are assessed simultaneously. The t-table is, furthermore, full of t-values, that, with ∞ degrees of freedom, are equal to z-values (Chap. 36). The t-values are to be understood as mean results of studies, but not expressed in mmol/l, kilograms, but in so-called SEM-units (Standard error of the mean units), that are obtained by dividing your mean result by its own standard error. With many degrees of freedom (large samples) the curve will be a little bit narrower, and more in agreement with nature.

The ninth row of t-values shows that our t-value is between 2.262 and 2.821. This would mean a p-value between 0.05 and 0.02. There is, thus, a significant difference between observation 1 and 2. However, is the expected power obtained or is this study underpowered. The t-table is helpful to calculate the t-value required for a power of 90 %: it mean a beta-value (type II error value) of 10 % (= 0.1). Look at the upper row of the t-table.

If
beta = 0.1, then
$z_{beta}$ for 9 degrees of freedom
= 1.383.

The t-value required for a power of 90 %

$= 1.383 + t^1$, where $t^1$ is the 0.05
$= 1.383 + 2.262$
$= 3.645.$

The required t-value is much larger than the obtained t-value of 2.809, and, so, the study does not meet its expected power. The treatment is less efficaceous than expected.

If the investigators had required a power of 60 %, then the superiority testing would be as follows.

beta = 0.40
z = 0.261

The t-value required for a power of 60 %

$= 0.261 + t^1$, where $t^1$ is the 0.05
$= 0.261 + 2.262$
$= 2.523.$

This t-value is smaller than the obtained t-value of 2.809, and, so, the study would have met an expected power of 60 %.

# 6 Conclusion

Superiority testing of a study means testing whether the study meets its a priori defined expected power. Many therapeutic studies may be able to reject their null-hypotheses, and, are, thus, statistically significant, but they do not meet their expected power. Superiority testing assesses whether the eventual power of a study is in agreement with the power as stated in the sample size calculation of the study. This chapter shows that with the help of the t-table the presence of superiority can be readily assessed.

We should note that the terms z-value and t-values are often used interchangeably, but strictly the z-value is the test statistic of the z-test, and the t-value is the test statistic of the t-test. The bottom row of the t-able is equal to the z-table.

## 7 Note

More background, theoretical and mathematical information, and alternative approaches to superiority testing is given in Statistics applied to clinical studies 5th edition, Chap. 62, Springer Heidelberg Germany, 2012, from the same authors.

# Chapter 17
# Missing Data Imputation

## 1 General Purpose

Missing data in clinical research data is often a real problem. As an example, a 35 patient data file of 3 variables consists of 3 × 35 = 105 values if the data are complete. With only 5 values missing (1 value missing per patient) 5 patients will not have complete data, and are rather useless for the analysis. This is not 5 % but 15 % of this small study population of 35 patients. An analysis of the remaining 85 % patients is likely not to be powerful to demonstrate the effects we wished to assess. This illustrates the necessity of data imputation.

## 2 Schematic Overview of Type of Data File

| Outcome |
|---|
| • |
| • |
| • |
| • |
| • |
| • |
| • |
| • |
| • |
| |
| |
| |
| |
| |



T.J. Cleophas, A.H. Zwinderman, *Clinical Data Analysis on a Pocket Calculator*,
DOI 10.1007/978-3-319-27104-0_17

## 3 Primary Scientific Question

Is mean and hot deck imputation capable of improving sensitivity of testing data files by increasing their fit to some analytical model.

## 4 Data Example

Four methods of data imputation are available; (1) mean imputation, (2) hot deck imputation, (3) regression imputation, (4) multiple imputations. In the current chapter the methods (1) and (2) will be given. The methods (3) and (4) are described in Statistics applied to clinical studies 5th edition, Chap. 22, Springer Heidelberg Germany, 2012, from the same authors as the current work. A condition for any type of data imputation is that the missing data are not clustered but randomly distributed in the data file. A data example a 35 patient study crossover study of the effects of age and traditional laxative efficacy (numbers of stools per month) on the performance of a novel laxative is in the underneath table.

| New lax | Bisacodyl | Age |
|---|---|---|
| 24,00 | 8,00 | 25,00 |
| 30,00 | 13,00 | 30,00 |
| 25,00 | 15,00 | 25,00 |
| 35,00 | 10,00 | 31,00 |
| 39,00 | 9,00 | |
| 30,00 | 10,00 | 33,00 |
| 27,00 | 8,00 | 22,00 |
| 14,00 | 5,00 | 18,00 |
| 39,00 | 13,00 | 14,00 |
| 42,00 | | 30,00 |
| 41,00 | 11,00 | 36,00 |
| 38,00 | 11,00 | 30,00 |
| 39,00 | 12,00 | 27,00 |
| 37,00 | 10,00 | 38,00 |
| 47,00 | 18,00 | 40,00 |
| | 13,00 | 31,00 |
| 36,00 | 12,00 | 25,00 |
| 12,00 | 4,00 | 24,00 |
| 26,00 | 10,00 | 27,00 |
| 20,00 | 8,00 | 20,00 |
| 43,00 | 16,00 | 35,00 |
| 31,00 | 15,00 | 29,00 |
| 40,00 | 14,00 | 32,00 |
| 31,00 | | 30,00 |

(continued)

| New lax | Bisacodyl | Age |
|---|---|---|
| 36,00 | 12,00 | 40,00 |
| 21,00 | 6,00 | 31,00 |
| 44,00 | 19,00 | 41,00 |
| 11,00 | 5,00 | 26,00 |
| 27,00 | 8,00 | 24,00 |
| 24,00 | 9,00 | 30,00 |
| 40,00 | 15,00 | |
| 32,00 | 7,00 | 31,00 |
| 10,00 | 6,00 | 23,00 |
| 37,00 | 14,00 | 43,00 |
| 19,00 | 7,00 | 30,00 |

Five values of the above table are missing. The underneath table gives the results of mean imputation of these data. The missing values are imputed by the mean values of the different variables.

| New lax | Bisacodyl | Age |
|---|---|---|
| 24,00 | 8,00 | 25,00 |
| 30,00 | 13,00 | 30,00 |
| 25,00 | 15,00 | 25,00 |
| 35,00 | 10,00 | 31,00 |
| 39,00 | 9,00 | *29,00* |
| 30,00 | 10,00 | 33,00 |
| 27,00 | 8,00 | 22,00 |
| 14,00 | 5,00 | 18,00 |
| 39,00 | 13,00 | 14,00 |
| 42,00 | *11,00* | 30,00 |
| 41,00 | 11,00 | 36,00 |
| 38,00 | 11,00 | 30,00 |
| 39,00 | 12,00 | 27,00 |
| 37,00 | 10,00 | 38,00 |
| 47,00 | 18,00 | 40,00 |
| *30,00* | 13,00 | 31,00 |
| 36,00 | 12,00 | 25,00 |
| 12,00 | 4,00 | 24,00 |
| 26,00 | 10,00 | 27,00 |
| 20,00 | 8,00 | 20,00 |
| 43,00 | 16,00 | 35,00 |
| 31,00 | 15,00 | 29,00 |
| 40,00 | 14,00 | 32,00 |
| 31,00 | *11,00* | 30,00 |
| 36,00 | 12,00 | 40,00 |
| 21,00 | 6,00 | 31,00 |
| 44,00 | 19,00 | 41,00 |

(continued)

| New lax | Bisacodyl | Age |
|---|---|---|
| 11,00 | 5,00 | 26,00 |
| 27,00 | 8,00 | 24,00 |
| 24,00 | 9,00 | 30,00 |
| 40,00 | 15,00 | *29,00* |
| 32,00 | 7,00 | 31,00 |
| 10,00 | 6,00 | 23,00 |
| 37,00 | 14,00 | 43,00 |
| 19,00 | 7,00 | 30,00 |

The underneath table gives the results of the second method, hot deck imputation. The missing data are imputed by those of the closest neighbour observed: the missing value is imputed with the value of an individual whose non-missing data are closest to those of the patient with the missing value.

| New lax | Bisacodyl | Age |
|---|---|---|
| 24,00 | 8,00 | 25,00 |
| 30,00 | 13,00 | 30,00 |
| 25,00 | 15,00 | 25,00 |
| 35,00 | 10,00 | 31,00 |
| 39,00 | 9,00 | *30,00* |
| 30,00 | 10,00 | 33,00 |
| 27,00 | 8,00 | 22,00 |
| 14,00 | 5,00 | 18,00 |
| 39,00 | 13,00 | 14,00 |
| 42,00 | *14,00* | 30,00 |
| 41,00 | 11,00 | 36,00 |
| 38,00 | 11,00 | 30,00 |
| 39,00 | 12,00 | 27,00 |
| 37,00 | 10,00 | 38,00 |
| 47,00 | 18,00 | 40,00 |
| *30,00* | 13,00 | 31,00 |
| 36,00 | 12,00 | 25,00 |
| 12,00 | 4,00 | 24,00 |
| 26,00 | 10,00 | 27,00 |
| 20,00 | 8,00 | 20,00 |
| 43,00 | 16,00 | 35,00 |
| 31,00 | 15,00 | 29,00 |
| 40,00 | 14,00 | 32,00 |
| 31,00 | *15,00* | 30,00 |
| 36,00 | 12,00 | 40,00 |
| 21,00 | 6,00 | 31,00 |
| 44,00 | 19,00 | 41,00 |
| 11,00 | 5,00 | 26,00 |
| 27,00 | 8,00 | 24,00 |

(continued)

| New lax | Bisacodyl | Age |
|---|---|---|
| 24,00 | 9,00 | 30,00 |
| 40,00 | 15,00 | *32,00* |
| 32,00 | 7,00 | 31,00 |
| 10,00 | 6,00 | 23,00 |
| 37,00 | 14,00 | 43,00 |
| 19,00 | 7,00 | 30,00 |

# 5 Conclusion

Imputed data are of course not real data, but constructed values that should increase the sensitivity of testing. Regression imputation is more sensitive than mean and hot deck imputation, but it often overstates sensitivity. Probably, the best method for data imputation is multiple imputations (4), because this method works as a device for representing missing data uncertainty. However, a pocket calculator is unable to perform the analysis, and a statistical software package like SPSS statistical software is required.

# 6 Note

More background, theoretical and mathematical information of missing data and data imputations is given in Statistics applied to clinical studies 5th edition, Chap. 22, Springer Heidelberg Germany, 2012, from the same authors.

# Chapter 18
# Bonferroni Adjustments

## 1 General Purpose

The unpaired t-test can be used to test the hypothesis that the means of two parallel-group are not different (Chap. 7). When the experimental design involves multiple groups, and, thus, multiple tests, it will increase the chance of finding differences. This is, simply, due to the play of chance, rather than a real effect. Multiple testing without any adjustment for this increased chance is called data dredging, and is the source of multiple type I errors (chances of finding a difference where there is none). The Bonferroni adjusted t-test (and many other methods) are appropriate for adjusting the increased risk of type I errors. This chapter will assess how it works.

## 2 Schematic Overview of Type of Data File

| Group (1,2,3,...) | Outcome |
|---|---|
| . | |
| . | |
| . | |
| . | |
| . | |
| . | |
| . | |
| . | |
| . | |
| | |
| | |
| | |
| | |



T.J. Cleophas, A.H. Zwinderman, *Clinical Data Analysis on a Pocket Calculator*,
DOI 10.1007/978-3-319-27104-0_18

## 3 Primary Scientific Question

In a parallel-group study of three or more treatments, does Bonferroni t-test adequately adjust the increased risk of type I errors?

## 4 Bonferroni T-Test, Data Example

The underneath example studies three groups of patients treated with different hemoglobin improving compounds. The mean increases of hemoglobin are given.

| | Sample size | Mean hemoglobin mmol/l | Standard deviation mmol/l |
|---|---|---|---|
| Group1 | 16 | 8.725 | 0.8445 |
| Group 2 | 10 | 10.6300 | 1.2841 |
| Group 3 | 15 | 12.3000 | 0.9419 |

An overall analysis of variance test produced a p-value of $< 0.01$. The conclusion is that we have a significant difference in the data, but we will need additional testing to find out, exactly where the difference is:

between group 1 and 2,
between group 1 and 3, or
between group 2 and 3.

The easiest way is to perform a t-test for each comparison. It produces a highly significant difference at $p < 0.01$ between group 1 versus 3 with no significant differences between the other comparisons. This highly significant result is, however, unadjusted for multiple comparisons. If one analyzes a set of data with three t-tests, each using a 5 % critical value for concluding that there is a significant difference, then there is about $3 \times 5 = 15$ % chance of finding a significant difference at least once. This mechanism is called the Bonferroni inequality. Bonferroni recommended a solution for the inequality, and proposed to follow in case of three t-tests to use a smaller critical level for concluding that there is a significant difference:

With 1t-test: critical level = 5 %
With 2t-tests: critical level = (5 %) /2 = 2,5 %
With 3t-tests: critical level = (5 %) /3 = 1.67 %.

a more general version of the equation is given underneath:

In case of k comparisons and an overall critical level (= null-hypothesis rejection level) of $\alpha$ the rejection p-value will become

$$\alpha \times 2/(k(k-1))$$

E.g. with $k = 3$, and $\alpha = 0.05$ (5 %)

$$0.05 \times \frac{2}{3(3-1)} = 0.0166.$$

In the given example a p-value of 0.0166 is still larger than 0.01, and, so, the difference observed remained statistically significant, but using a cut-off p-value of 0.0166, instead of 0.05, means that the difference is not *highly* significant anymore.

## 5 Bonferroni T-Test Over-conservative

With k-values not too large, this method performs well. However, if k is large ($k > 5$), then this Bonferroni correction will rapidly lead to very small rejection p-values, and is, therefore, called over-conservative. E.g., with 10 comparisons the rejection p-value will be only $0.05 \times 2/(10 \times 9) = 0.001$, which is a value hard to be obtained in a small study. Actually, a more realistic rejection p-value may be larger, because in most multiple test studies a positive correlation between multiple treatment comparisons exists. It means that multiple tests in one study are much more at risk of similar results than multiple tests in different studies (see also Chap. 10). The chance of twice a p-value of 0.05 may, therefore, not be 0.025, but, rather, something in between, like 0.035.

## 6 Conclusion

Bonferroni adjustment is adequate for adjusting the increased type I error of multiple testing, and can easily be performed with the help of a pocket calculator. Alternative methods include Tukey's honestly significant difference (HSD) method, Student-Newman-Keuls method, the method of Dunnett, and many more. They are, however, computationally more laborious, and require specific statistical tables. Statistical software programs like SPSS or SAS are helpful.

Bonferroni adjustments increase the risk of type II errors ($\beta$-values) of not finding a difference which does exist. This is illustrated in the underneath figure: with $\alpha = 0.05$ (left vertical interrupted line), $\beta$ is about 30 % of the area under the curve. With $\alpha = 0.167$ (adjusted for three tests as demonstrated in the above Sect. 4) (right vertical interrupted line), $\beta$ has risen to about 50 %. This rise caused loss of power from about 70 % to about only 50 % ($(1-\beta)$-values), (see also Chap. 11 for additional explanation of power assessments). H0 = null-hypothesis, H1 = alternative hypothesis, SEM = standard error of the mean.

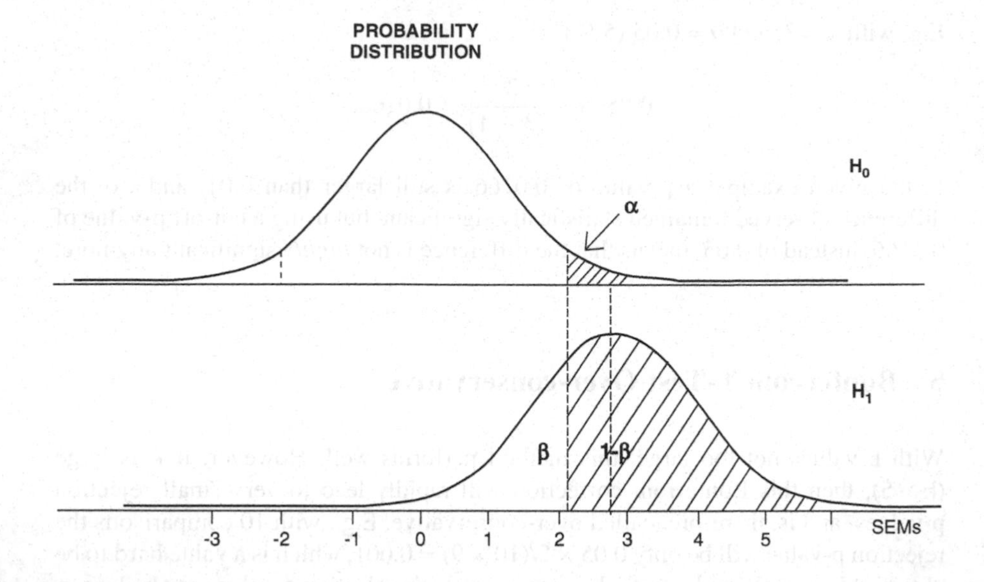


In the current chapter only continuous outcome data are adjusted for multiple testing. However, binary data can equally be assessed using the Bonferroni equation.

## 7 Note

More background, theoretical and mathematical information of multiple comparisons and false positive studies are given in Statistics applied to clinical studies 5th edition, Chaps. 8 and 9, Springer Heidelberg Germany, from the same authors.

# Chapter 19
# Unpaired Analysis of Variance

## 1 General Purpose

Unpaired t-tests are for assessing two parallel groups of similar age, gender and other characteristics treated differently. However, if you wish to compare three different treatments, three parallel groups are required, and unpaired t-tests can no longer be applied. Instead, unpaired analysis of variance (ANOVA), otherwise called one-way ANOVA must be used for analysis.

## 2 Schematic Overview of Type of Data File

| Outcome | Treatment group (1, 2, 3, ....) |
|---|---|
| . | . |
| . | . |
| . | . |
| . | . |
| . | . |
| . | . |
| . | . |
| . | . |
| . | . |
| | |
| | |
| | |
| | |

 
T.J. Cleophas, A.H. Zwinderman, *Clinical Data Analysis on a Pocket Calculator*,
DOI 10.1007/978-3-319-27104-0_19

## 3 Primary Scientific Question

How can one-way ANOVA evaluate the difference between three treatments.

## 4 Variations Expressed as the Sums of Squares

With unpaired ANOVA of 3 treatment groups variations between the data are split:

Total variation
| |
Between group variation within group variation

Variations is expressed as the sums of squares (SS) and can be added up to obtain total variation. We wish to assess whether the between-group variation is large compared to the within-group variation. This approach may be hard to understand, but the underneath simple example shows the essentials of it (n = number of patients, SD = standard deviation).

| Group | n patients | mean | SD |
|---|---|---|---|
| 1 | n | – | – |
| 2 | n | – | – |
| 3 | n | – | – |

Grand mean = (mean 1 + 2 + 3) / 3

SS beween-groups = n $(\text{mean}_1 - \text{grand mean})^2$ + n $(\text{mean}_2 - \text{grand mean})^2$ + ....
SS within-groups = $(n-1)(SD_1{}^2) + (n-1)(SD_2{}^2) + \ldots$..

The F-test (Fisher-test) is used for testing (dfs = degrees of freedom):

$$F = \frac{\text{SS between-groups/dfs}}{\text{SS within-groups/dfs}} = \frac{\text{SS between-groups}/(3-1)}{\text{SS within-groups}/(3n-3).}$$

The F-table gives the p-value.

We should note, that, with differently sized groups, weighted grand means are required: weighted mean = $(n_1 \text{ mean}_1 + n_2 \text{ mean}_2) / (n_1 + n_2)$.

## 5 Real Data Example

Effect of 3 compounds on Hb

| Group | n patients | mean | SD |
|---|---|---|---|
| 1 | 16 | 8.7125 | 0.8445 |
| 2 | 16 | 10.6300 | 1.2841 |
| 3 | 16 | 12.3000 | 0.9419 |

Grand mean = (mean 1 + 2 + 3) / 3 = 10.4926

SS between-groups = $16(8.7125 - 10.4926)^2 + 16(10.6300 - 10.4926)^2 \ldots$

SS within-groups $= 15 \times 0.84452 + 15 \times 1.28412 + \ldots\ldots$

$$F = 49.9$$

| *df* of | 2-tailed | 1-tailed | Degrees of freedom (*df*) of the numerator | | | | | | | | | | | | |
|---|---|---|---|---|---|---|---|---|---|---|---|---|---|---|---|
| denominator | *P*-value | *P*-value | 1 | 2 | 3 | 4 | 5 | 6 | 7 | 8 | 9 | 10 | 15 | 25 | 500 |
| 1 | 0.05 | 0.025 | 647.8 | 799.5 | 864.2 | 899.6 | 921.8 | 937.1 | 948.2 | 956.6 | 963.3 | 968.6 | 984.9 | 998.1 | 1017.0 |
| 1 | 0.10 | 0.05 | 161.4 | 199.5 | 215.7 | 224.6 | 230.2 | 234.0 | 236.8 | 238.9 | 240.5 | 241.9 | 245.9 | 249.3 | 254.1 |
| 2 | 0.05 | 0.025 | 38.51 | 39.00 | 39.17 | 39.25 | 39.30 | 39.33 | 39.36 | 39.37 | 39.39 | 39.40 | 39.43 | 39.46 | 39.50 |
| 2 | 0.10 | 0.05 | 18.51 | 19.00 | 19.16 | 19.25 | 19.30 | 19.33 | 19.35 | 19.37 | 19.38 | 19.40 | 19.43 | 19.46 | 19.49 |
| 3 | 0.05 | 0.025 | 17.44 | 16.04 | 15.44 | 15.10 | 14.88 | 14.73 | 14.62 | 14.54 | 14.47 | 14.42 | 14.25 | 14.12 | 13.91 |
| 3 | 0.10 | 0.05 | 10.13 | 9.55 | 9.28 | 9.12 | 9.01 | 8.94 | 8.89 | 8.85 | 8.81 | 8.79 | 8.70 | 8.63 | 8.53 |
| 4 | 0.05 | 0.025 | 12.22 | 10.65 | 9.98 | 9.60 | 9.36 | 9.20 | 9.07 | 8.98 | 8.90 | 8.84 | 8.66 | 8.50 | 8.27 |
| 4 | 0.10 | 0.05 | 7.71 | 6.94 | 6.59 | 6.39 | 6.26 | 6.16 | 6.09 | 6.04 | 6.00 | 5.96 | 5.86 | 5.77 | 5.64 |
| 5 | 0.05 | 0.025 | 10.01 | 8.43 | 7.76 | 7.39 | 7.15 | 6.98 | 6.85 | 6.76 | 6.68 | 6.62 | 6.43 | 6.27 | 6.03 |
| 5 | 0.10 | 0.05 | 6.61 | 5.79 | 5.41 | 5.19 | 5.05 | 4.95 | 4.88 | 4.82 | 4.77 | 4.74 | 4.62 | 4.52 | 4.37 |
| 6 | 0.05 | 0.025 | 8.81 | 7.26 | 6.60 | 6.23 | 5.99 | 5.82 | 5.70 | 5.60 | 5.52 | 5.46 | 5.27 | 5.11 | 4.86 |
| 6 | 0.10 | 0.05 | 5.99 | 5.14 | 4.76 | 4.53 | 4.39 | 4.28 | 4.21 | 4.15 | 4.10 | 4.06 | 3.94 | 3.83 | 3.68 |
| 7 | 0.05 | 0.025 | 8.07 | 6.54 | 5.89 | 5.52 | 5.29 | 5.12 | 4.99 | 4.90 | 4.82 | 4.76 | 4.57 | 4.40 | 4.16 |
| 7 | 0.10 | 0.05 | 5.59 | 4.74 | 4.35 | 4.12 | 3.97 | 3.87 | 3.79 | 3.73 | 3.68 | 3.64 | 3.51 | 3.40 | 3.24 |
| 8 | 0.05 | 0.025 | 7.57 | 6.06 | 5.42 | 5.05 | 4.82 | 4.65 | 4.53 | 4.43 | 4.36 | 4.30 | 4.10 | 3.94 | 3.68 |
| 8 | 0.10 | 0.05 | 5.32 | 4.46 | 4.07 | 3.84 | 3.69 | 3.58 | 3.50 | 3.44 | 3.39 | 3.35 | 3.22 | 3.11 | 2.94 |
| 9 | 0.05 | 0.025 | 7.21 | 5.71 | 5.08 | 4.72 | 4.48 | 4.32 | 4.20 | 4.10 | 4.03 | 3.96 | 3.77 | 3.60 | 3.35 |
| 9 | 0.10 | 0.05 | 5.12 | 4.26 | 3.86 | 3.63 | 3.48 | 3.37 | 3.29 | 3.23 | 3.18 | 3.14 | 3.01 | 2.89 | 2.72 |
| 10 | 0.05 | 0.025 | 6.94 | 5.46 | 4.83 | 4.47 | 4.24 | 4.07 | 3.95 | 3.85 | 3.78 | 3.72 | 3.52 | 3.35 | 3.09 |
| 10 | 0.10 | 0.05 | 4.96 | 4.10 | 3.71 | 3.48 | 3.33 | 3.22 | 3.14 | 3.07 | 3.02 | 2.98 | 2.85 | 2.73 | 2.55 |
| 15 | 0.05 | 0.025 | 6.20 | 4.77 | 4.15 | 3.80 | 3.58 | 3.41 | 3.29 | 3.20 | 3.12 | 3.06 | 2.86 | 2.69 | 2.41 |
| 15 | 0.10 | 0.05 | 4.54 | 3.68 | 3.29 | 3.06 | 2.90 | 2.79 | 2.71 | 2.64 | 2.59 | 2.54 | 2.40 | 2.28 | 2.08 |
| 20 | 0.05 | 0.025 | 5.87 | 4.46 | 3.86 | 3.51 | 3.29 | 3.13 | 3.01 | 2.91 | 2.84 | 2.77 | 2.57 | 2.40 | 2.10 |
| 20 | 0.10 | 0.05 | 4.35 | 3.49 | 3.10 | 2.87 | 2.71 | 2.60 | 2.51 | 2.45 | 2.39 | 2.35 | 2.20 | 2.07 | 1.86 |
| 30 | 0.05 | 0.025 | 5.57 | 4.18 | 3.59 | 3.25 | 3.03 | 2.87 | 2.75 | 2.65 | 2.57 | 2.51 | 2.31 | 2.12 | 1.81 |
| 30 | 0.10 | 0.05 | 4.17 | 3.32 | 2.92 | 2.69 | 2.53 | 2.42 | 2.33 | 2.27 | 2.21 | 2.16 | 2.01 | 1.88 | 1.64 |
| 50 | 0.05 | 0.025 | 5.34 | 3.97 | 3.39 | 3.05 | 2.83 | 2.67 | 2.55 | 2.46 | 2.38 | 2.32 | 2.11 | 1.92 | 1.57 |
| 50 | 0.10 | 0.05 | 4.03 | 3.18 | 2.79 | 2.56 | 2.40 | 2.29 | 2.20 | 2.13 | 2.07 | 2.03 | 1.87 | 1.73 | 1.46 |
| 100 | 0.05 | 0.025 | 5.18 | 3.83 | 3.25 | 2.92 | 2.70 | 2.54 | 2.42 | 2.32 | 2.24 | 2.18 | 1.97 | 1.77 | 1.38 |
| 100 | 0.10 | 0.05 | 3.94 | 3.09 | 2.70 | 2.46 | 2.31 | 2.19 | 2.10 | 2.03 | 1.97 | 1.93 | 1.77 | 1.62 | 1.31 |
| 1000 | 0.05 | 0.025 | 5.04 | 3.70 | 3.13 | 2.80 | 2.58 | 2.42 | 2.30 | 2.20 | 2.13 | 2.06 | 1.85 | 1.64 | 1.16 |
| 1000 | 0.10 | 0.05 | 3.85 | 3.00 | 2.61 | 2.38 | 2.22 | 2.11 | 2.02 | 1.95 | 1.89 | 1.84 | 1.68 | 1.52 | 1.13 |

This value is much larger than the critical F-value producing a $p < 0.05$, because, with 2 (numerator) and 45 (denominator) degrees of freedom the critical F-value should be between 3.32 and 3.97. The difference between the three treatment is, thus, very significant. A table of critical F-values is given on the next page. The internet provides, however, many critical F-value calculators, that are more precise.

We should add that, in case 2 groups, the F-value produced by ANOVA equals the t-test squared ($F = t^2$). T-statistics is, indeed, a simple form of analysis of variance.

## 6 Conclusion

The above examples show how one-way ANOVA can be used to test the significance of difference between three treatments. However, it does not tell us whether treatment 1 is better than 2, 2 better than 3, or 1 better than 3, or any combinations of these effects. For that purpose post hoc tests are required comparing the treatments one by one. Unpaired t-tests should be appropriate for the purpose.

## 7 Note

More background, theoretical and mathematical information of unpaired and paired ANOVA is given Statistics applied to clinical studies 5th edition, Chap. 2, Springer Heidelberg Germany, 2012, from the same authors.

# Chapter 20
# Paired Analysis of Variance

## 1 General Purpose

Paired t-tests are for assessing the effect of two treatments in a single group of patients. However, if you wish to compare three treatments, the paired t-tests can no longer be applied. Instead, paired analysis of variance (ANOVA), otherwise called repeated-measures ANOVA must be used for analysis.

## 2 Schematic Overview of Type of Data File

| | Outcome | Outcome | Outcome |
|---|---|---|---|
| Patient no | Treatment 1 | Treatment 2 | Treatment 3 |
| . | . | . | . |
| . | . | . | . |
| . | . | . | . |
| . | . | . | . |
| . | . | . | . |
| . | . | . | . |
| . | . | . | . |
| . | . | . | . |
| . | . | . | . |
| . | . | . | . |
| | | | |
| | | | |
| | | | |
| | | | |
| | | | |



T.J. Cleophas, A.H. Zwinderman, *Clinical Data Analysis on a Pocket Calculator*,
DOI 10.1007/978-3-319-27104-0_20

## 3 Primary Scientific Question

How can repeated-measures ANOVA evaluate the different effect of three different treatments.

## 4 Variations Expressed as the Sums of Squares

With paired ANOVA of three treatments every single patient is treated three times. The data are split:

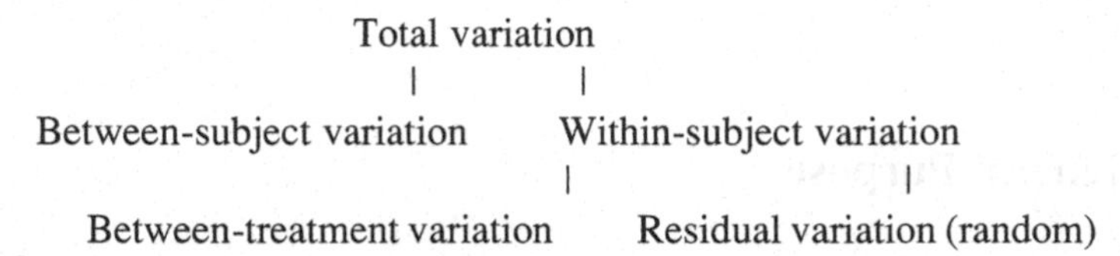


Variations is expressed as the sums of squares (SS), and can be added up to obtain the total variation in the data. We will assess, whether the between-treatment variation is large compared to the residual variation. Repeated-measure ANOVA is sometimes called two-way ANOVA (balanced, without replications), (SD = standard deviation).

| Subject | Treatment 1 | Treatment 2 | Treatment 3 | $SD^2$ |
|---|---|---|---|---|
| 1 | – | – | – | – |
| 2 | – | – | – | – |
| 3 | – | – | – | – |
| 4 | – | – | – | – |
| Treatment mean | – | – | – | |

Grand mean = (treatment mean 1 + 2 + 3) / 3 = .....

SS within-subject = $SD_1{}^2 + SD_2{}^2 + SD_3{}^2$

SS treatment = (treatment mean 1 − grand mean)$^2$ + (treatment mean 2 − grand mean)$^2$ + .....

SS residual = SS within-subject − SS treatment

The F-test (Fisher-test) is used for testing (dfs = degrees of freedom):

$$F = \frac{\text{SS treatment /dfs}}{\text{SS residual / dfs}} = \frac{\text{SS treatment /}(3-1)}{\text{SS residual / }(3-1)(4-1)}$$

The F-table gives the P-value.

# 5 Real Data Example

The effect of 3 treatments on vascular resistance is assessed in four persons.

| Person | Treatment 1 | Treatment 2 | Treatment 3 | $SD^2$ |
|---|---|---|---|---|
| 1 | 22.2 | 5.4 | 10.6 | 147.95 |
| 2 | 17.0 | 6.3 | 6.2 | 77.05 |
| 3 | 14.1 | 8.5 | 9.3 | 18.35 |
| 4 | 17.0 | 10.7 | 12.3 | 21.45 |
| Treatment mean | 17.58 | 7.73 | 9.60 | |

Grand mean = 11.63

SS within-subject = 147.95 + 77.05 + ..
SS treatment = $(17.58 - 11.63)^2 + (7.73 - 11.63)^2 + ..$
SS residual = SS within-subject – SS treatment.
F = 14.31.

This value is much larger than the critical F-value producing a $p < 0.05$, because with 2 (numerator) degrees of freedom and $2 \times 3 = 6$ (denominator) degrees of freedom the critical F-value should be around 7.26. The difference between the effects of the three treatments is, thus, very significant. The table of critical F-values is given on the next page. The internet provides, however, many critical F-value calculators, that are more precise than the table given.

We should add that, in case of 2 treatments the F-value produced by the ANOVA equals the t-value squared ($F = t^2$). T-statistics is, indeed, a simple form of analysis of variance.

| *df* of denominator | 2-tailed *P*-value | 1-tailed *P*-value | Degrees of freedom (*df*) of the numerator | | | | | | | | | | | | |
|---|---|---|---|---|---|---|---|---|---|---|---|---|---|---|---|
| | | | 1 | 2 | 3 | 4 | 5 | 6 | 7 | 8 | 9 | 10 | 15 | 25 | 500 |
| 1 | 0.05 | 0.025 | 647.8 | 799.5 | 864.2 | 899.6 | 921.8 | 937.1 | 948.2 | 956.6 | 963.3 | 968.6 | 984.9 | 998.1 | 1017.0 |
| 1 | 0.10 | 0.05 | 161.4 | 199.5 | 215.7 | 224.6 | 230.2 | 234.0 | 236.8 | 238.9 | 240.5 | 241.9 | 245.9 | 249.3 | 254.1 |
| 2 | 0.05 | 0.025 | 38.51 | 39.00 | 39.17 | 39.25 | 39.30 | 39.33 | 39.36 | 39.37 | 39.39 | 39.40 | 39.43 | 39.46 | 39.50 |
| 2 | 0.10 | 0.05 | 18.51 | 19.00 | 19.16 | 19.25 | 19.30 | 19.33 | 19.35 | 19.37 | 19.38 | 19.40 | 19.43 | 19.46 | 19.49 |
| 3 | 0.05 | 0.025 | 17.44 | 16.04 | 15.44 | 15.10 | 14.88 | 14.73 | 14.62 | 14.54 | 14.47 | 14.42 | 14.25 | 14.12 | 13.91 |
| 3 | 0.10 | 0.05 | 10.13 | 9.55 | 9.28 | 9.12 | 9.01 | 8.94 | 8.89 | 8.85 | 8.81 | 8.79 | 8.70 | 8.63 | 8.53 |
| 4 | 0.05 | 0.025 | 12.22 | 10.65 | 9.98 | 9.60 | 9.36 | 9.20 | 9.07 | 8.98 | 8.90 | 8.84 | 8.66 | 8.50 | 8.27 |
| 4 | 0.10 | 0.05 | 7.71 | 6.94 | 6.59 | 6.39 | 6.26 | 6.16 | 6.09 | 6.04 | 6.00 | 5.96 | 5.86 | 5.77 | 5.64 |
| 5 | 0.05 | 0.025 | 10.01 | 8.43 | 7.76 | 7.39 | 7.15 | 6.98 | 6.85 | 6.76 | 6.68 | 6.62 | 6.43 | 6.27 | 6.03 |
| 5 | 0.10 | 0.05 | 6.61 | 5.79 | 5.41 | 5.19 | 5.05 | 4.95 | 4.88 | 4.82 | 4.77 | 4.74 | 4.62 | 4.52 | 4.37 |
| 6 | 0.05 | 0.025 | 8.81 | 7.26 | 6.60 | 6.23 | 5.99 | 5.82 | 5.70 | 5.60 | 5.52 | 5.46 | 5.27 | 5.11 | 4.86 |
| 6 | 0.10 | 0.05 | 5.99 | 5.14 | 4.76 | 4.53 | 4.39 | 4.28 | 4.21 | 4.15 | 4.10 | 4.06 | 3.94 | 3.83 | 3.68 |
| 7 | 0.05 | 0.025 | 8.07 | 6.54 | 5.89 | 5.52 | 5.29 | 5.12 | 4.99 | 4.90 | 4.82 | 4.76 | 4.57 | 4.40 | 4.16 |
| 7 | 0.10 | 0.05 | 5.59 | 4.74 | 4.35 | 4.12 | 3.97 | 3.87 | 3.79 | 3.73 | 3.68 | 3.64 | 3.51 | 3.40 | 3.24 |
| 8 | 0.05 | 0.025 | 7.57 | 6.06 | 5.42 | 5.05 | 4.82 | 4.65 | 4.53 | 4.43 | 4.36 | 4.30 | 4.10 | 3.94 | 3.68 |
| 8 | 0.10 | 0.05 | 5.32 | 4.46 | 4.07 | 3.84 | 3.69 | 3.58 | 3.50 | 3.44 | 3.39 | 3.35 | 3.22 | 3.11 | 2.94 |
| 9 | 0.05 | 0.025 | 7.21 | 5.71 | 5.08 | 4.72 | 4.48 | 4.32 | 4.20 | 4.10 | 4.03 | 3.96 | 3.77 | 3.60 | 3.35 |
| 9 | 0.10 | 0.05 | 5.12 | 4.26 | 3.86 | 3.63 | 3.48 | 3.37 | 3.29 | 3.23 | 3.18 | 3.14 | 3.01 | 2.89 | 2.72 |
| 10 | 0.05 | 0.025 | 6.94 | 5.46 | 4.83 | 4.47 | 4.24 | 4.07 | 3.95 | 3.85 | 3.78 | 3.72 | 3.52 | 3.35 | 3.09 |
| 10 | 0.10 | 0.05 | 4.96 | 4.10 | 3.71 | 3.48 | 3.33 | 3.22 | 3.14 | 3.07 | 3.02 | 2.98 | 2.85 | 2.73 | 2.55 |
| 15 | 0.05 | 0.025 | 6.20 | 4.77 | 4.15 | 3.80 | 3.58 | 3.41 | 3.29 | 3.20 | 3.12 | 3.06 | 2.86 | 2.69 | 2.41 |
| 15 | 0.10 | 0.05 | 4.54 | 3.68 | 3.29 | 3.06 | 2.90 | 2.79 | 2.71 | 2.64 | 2.59 | 2.54 | 2.40 | 2.28 | 2.08 |
| 20 | 0.05 | 0.025 | 5.87 | 4.46 | 3.86 | 3.51 | 3.29 | 3.13 | 3.01 | 2.91 | 2.84 | 2.77 | 2.57 | 2.40 | 2.10 |
| 20 | 0.10 | 0.05 | 4.35 | 3.49 | 3.10 | 2.87 | 2.71 | 2.60 | 2.51 | 2.45 | 2.39 | 2.35 | 2.20 | 2.07 | 1.86 |
| 30 | 0.05 | 0.025 | 5.57 | 4.18 | 3.59 | 3.25 | 3.03 | 2.87 | 2.75 | 2.65 | 2.57 | 2.51 | 2.31 | 2.12 | 1.81 |
| 30 | 0.10 | 0.05 | 4.17 | 3.32 | 2.92 | 2.69 | 2.53 | 2.42 | 2.33 | 2.27 | 2.21 | 2.16 | 2.01 | 1.88 | 1.64 |
| 50 | 0.05 | 0.025 | 5.34 | 3.97 | 3.39 | 3.05 | 2.83 | 2.67 | 2.55 | 2.46 | 2.38 | 2.32 | 2.11 | 1.92 | 1.57 |
| 50 | 0.10 | 0.05 | 4.03 | 3.18 | 2.79 | 2.56 | 2.40 | 2.29 | 2.20 | 2.13 | 2.07 | 2.03 | 1.87 | 1.73 | 1.46 |
| 100 | 0.05 | 0.025 | 5.18 | 3.83 | 3.25 | 2.92 | 2.70 | 2.54 | 2.42 | 2.32 | 2.24 | 2.18 | 1.97 | 1.77 | 1.38 |
| 100 | 0.10 | 0.05 | 3.94 | 3.09 | 2.70 | 2.46 | 2.31 | 2.19 | 2.10 | 2.03 | 1.97 | 1.93 | 1.77 | 1.62 | 1.31 |
| 1000 | 0.05 | 0.025 | 5.04 | 3.70 | 3.13 | 2.80 | 2.58 | 2.42 | 2.30 | 2.20 | 2.13 | 2.06 | 1.85 | 1.64 | 1.16 |
| 1000 | 0.10 | 0.05 | 3.85 | 3.00 | 2.61 | 2.38 | 2.22 | 2.11 | 2.02 | 1.95 | 1.89 | 1.84 | 1.68 | 1.52 | 1.13 |

## 6 Conclusion

The above examples show how repeated-measures ANOVA can be used to test the significance of difference between three treatments in a single group of patients. However, it does not tell us whether treatment 1 is better than 2, treatment 2 better than 3, or treatment 1 better than 3, or any combinations of these effects. For that purpose post hoc tests are required, comparing the treatments one by one. Paired t-tests should be appropriate for the purpose.

## 7 Note

More background, theoretical and mathematical information of unpaired and paired ANOVA is given Statistics applied to clinical studies 5th edition, Chap. 2, Springer Heidelberg Germany, 2012, from the same authors.

# Chapter 21
# Variability Analysis for One or Two Samples

## 1 General Purpose

In some clinical studies, the spread of the data may be more relevant than the average of the data. E.g., when we assess how a drug reaches various organs, variability of drug concentrations is important, as in some cases too little and in other cases dangerously high levels get through. Also, variabilities in drug response may be important. For example, the spread of glucose levels of a slow-release-insulin is important. This chapter assesses how to estimate the spread of one and two data-samples.

## 2 Schematic Overview of Type of Data File

| Outcome |
|---|
| . |
| . |
| . |
| . |
| . |
| . |
| . |
| . |
| . |
| |
| |
| |
| |



T.J. Cleophas, A.H. Zwinderman, *Clinical Data Analysis on a Pocket Calculator*,
DOI 10.1007/978-3-319-27104-0_21

## 3 Primary Scientific Question

Is the spread in a data set larger than required, is the difference in variabilities between two data samples statistically significant.

## 4 One Sample Variability Analysis

For testing whether the standard deviation (or variance) of a sample is significantly different from the standard deviation (or variance) to be expected, the chi-square test with multiple degrees of freedom is adequate. The test statistic (the chi-square-value $= \chi^2$ –value) is calculated according to

$$\chi^2 = \frac{(n-1)s^2}{\sigma^2} \text{ for n} - 1 \text{ degrees of freedom}$$

n = sample size, s = standard deviation, $s^2$ = variance sample, σ = expected standard deviation, $\sigma^2$ = expected variance).

For example, the aminoglycoside compound gentamicin has a small therapeutic index. The standard deviation of 50 measurements is used as a criterion for variability. Adequate variability will be accepted, if the standard deviation is less than 7 μg/l. In our sample a standard deviation of 9 μg/l is observed. The test procedure is given.

$$\chi^2 = (50-1)\, 9^2/7^2 = 81$$

The underneath chi-square table has an upper row with areas under the curve, a left-end column with degrees of freedom, and a whole lot of chi-square values. It shows that, for 50–1 = 49 degrees of freedom (close to 50 df row), we will find that a chi-square value 76.154 will produce a p-value < 0.01. This sample's standard deviation is significantly larger than that required. This means that the variability in plasma gentamicin concentrations is larger than acceptable.

Chi-squared distribution

| | Two-tailed *P*-value | | | |
|---|---|---|---|---|
| *df* | 0.10 | 0.05 | 0.01 | 0.001 |
| 1 | 2.706 | 3.841 | 6.635 | 10.827 |
| 2 | 4.605 | 5.991 | 9.210 | 13.815 |
| 3 | 6.251 | 7.851 | 11.345 | 16.266 |
| 4 | 7.779 | 9.488 | 13.277 | 18.466 |
| 5 | 9.236 | 11.070 | 15.086 | 20.515 |

(continued)

| df | Two-tailed *P*-value | | | |
|---|---|---|---|---|
| | 0.10 | 0.05 | 0.01 | 0.001 |
| 6 | 10.645 | 12.592 | 16.812 | 22.457 |
| 7 | 12.017 | 14.067 | 18.475 | 24.321 |
| 8 | 13.362 | 15.507 | 20.090 | 26.124 |
| 9 | 14.684 | 16.919 | 21.666 | 27.877 |
| 10 | 15.987 | 18.307 | 23.209 | 29.588 |
| 11 | 17.275 | 19.675 | 24.725 | 31.264 |
| 12 | 18.549 | 21.026 | 26.217 | 32.909 |
| 13 | 19.812 | 22.362 | 27.688 | 34.527 |
| 14 | 21.064 | 23.685 | 29.141 | 36.124 |
| 15 | 22.307 | 24.996 | 30.578 | 37.698 |
| 16 | 23.542 | 26.296 | 32.000 | 39.252 |
| 17 | 24.769 | 27.587 | 33.409 | 40.791 |
| 18 | 25.989 | 28.869 | 34.805 | 42.312 |
| 19 | 27.204 | 30.144 | 36.191 | 43.819 |
| 20 | 28.412 | 31.410 | 37.566 | 45.314 |
| 21 | 29.615 | 32.671 | 38.932 | 46.796 |
| 22 | 30.813 | 33.924 | 40.289 | 48.268 |
| 23 | 32.007 | 35.172 | 41.638 | 49.728 |
| 24 | 33.196 | 36.415 | 42.980 | 51.179 |
| 25 | 34.382 | 37.652 | 44.314 | 52.619 |
| 26 | 35.536 | 38.885 | 45.642 | 54.051 |
| 27 | 36.741 | 40.113 | 46.963 | 55.475 |
| 28 | 37.916 | 41.337 | 48.278 | 56.892 |
| 29 | 39.087 | 42.557 | 49.588 | 58.301 |
| 30 | 40.256 | 43.773 | 50.892 | 59.702 |
| 40 | 51.805 | 55.758 | 63.691 | 73.403 |
| 50 | 63.167 | 67.505 | 76.154 | 86.660 |
| 60 | 74.397 | 79.082 | 88.379 | 99.608 |
| 70 | 85.527 | 90.531 | 100.43 | 112.32 |
| 80 | 96.578 | 101.88 | 112.33 | 124.84 |
| 90 | 107.57 | 113.15 | 124.12 | 137.21 |
| 100 | 118.50 | 124.34 | 135.81 | 149.45 |

## 5 Two Sample Variability Test

F-tests can be applied to test if the variabilities of two samples are significantly different from one another. The division sum of the samples' variances (larger variance / smaller variance) is used for the analysis. For example, two formulas of gentamicin produce the following standard deviations of plasma concentrations.

| | Patients (n) | Standard deviation (SD) (μg/l) |
|---|---|---|
| formula-A | 10 | 3.0 |
| formula-B | 15 | 2.0 |

$$\begin{aligned}\text{F-value} &= SD_A^2/SD_B^2\\ &= 3.0^2/2.0^2\\ &= 9/4 = 2.25\end{aligned}$$

with degrees of freedom (dfs) for

$$\begin{aligned}&\text{formula-A of } 10 - 1 = 9\\ &\text{formula-B of } 15 - 1 = 14\end{aligned}$$

The table of critical F-values producing a $p < 0.05$ is on the next page. It shows that with 9 and 14 degrees of freedom respectively in the numerator and denominator an F-value around 3.12 or more is required in order to reject the null – hypothesis. Our F-value is only 2.25, and, so, the p-value is $> 0.05$, and the null-hypothesis cannot be rejected. No significant difference between the two formulas can be demonstrated.

| *df* of denominator | 2-tailed | L-tailed | Degrees of freedom (*df*) of the numerator | | | | | | | | | | | | |
|---|---|---|---|---|---|---|---|---|---|---|---|---|---|---|---|
| | *P*-value | *P*-value | 1 | 2 | 3 | 4 | 5 | 6 | 7 | 8 | 9 | 10 | 15 | 25 | 500 |
| 1 | 0.05 | 0.025 | 647.8 | 799.5 | 864.2 | 899.6 | 921.8 | 937.1 | 948.2 | 956.6 | 963.3 | 968.6 | 984.9 | 998.1 | 1017.0 |
| 1 | 0.10 | 0.05 | 161.4 | 199.5 | 215.7 | 224.6 | 230.2 | 234.0 | 236.8 | 238.9 | 240.5 | 241.9 | 245.9 | 249.3 | 254.1 |
| 2 | 0.05 | 0.025 | 38.51 | 39.00 | 39.17 | 39.25 | 39.30 | 39.33 | 39.36 | 39.37 | 39.39 | 39.40 | 39.43 | 39.46 | 39.50 |
| 2 | 0.10 | 0.05 | 18.51 | 19.00 | 19.16 | 19.25 | 19.30 | 19.33 | 19.35 | 19.37 | 19.38 | 19.40 | 19.43 | 19.46 | 19.49 |
| 3 | 0.05 | 0.025 | 17.44 | 16.04 | 15.44 | 15.10 | 14.88 | 14.73 | 14.62 | 14.54 | 14.47 | 14.42 | 14.25 | 14.12 | 13.91 |
| 3 | 0.10 | 0.05 | 10.13 | 9.55 | 9.28 | 9.12 | 9.01 | 8.94 | 8.89 | 8.85 | 8.81 | 8.79 | 8.70 | 8.63 | 8.53 |
| 4 | 0.05 | 0.025 | 12.22 | 10.65 | 9.98 | 9.60 | 9.36 | 9.20 | 9.07 | 8.98 | 8.90 | 8.84 | 8.66 | 8.50 | 8.27 |
| 4 | 0.10 | 0.05 | 7.71 | 6.94 | 6.59 | 6.39 | 6.26 | 6.16 | 6.09 | 6.04 | 6.00 | 5.96 | 5.86 | 5.77 | 5.64 |
| 5 | 0.05 | 0.025 | 10.01 | 8.43 | 7.76 | 7.39 | 7.15 | 6.98 | 6.85 | 6.76 | 6.68 | 6.62 | 6.43 | 6.27 | 6.03 |
| 5 | 0.10 | 0.05 | 6.61 | 5.79 | 5.41 | 5.19 | 5.05 | 4.95 | 4.88 | 4.82 | 4.77 | 4.74 | 4.62 | 4.52 | 4.37 |
| 6 | 0.05 | 0.25 | 8.81 | 7.26 | 6.60 | 6.23 | 5.99 | 5.82 | 5.70 | 5.60 | 5.52 | 5.46 | 5.27 | 5.11 | 4.86 |
| 6 | 0.10 | 0.05 | 5.99 | 5.14 | 4.76 | 4.53 | 4.39 | 4.28 | 4.21 | 4.15 | 4.10 | 4.06 | 3.94 | 3.83 | 3.68 |
| 7 | 0.05 | 0.025 | 8.07 | 6.54 | 5.89 | 5.52 | 5.29 | 5.12 | 4.99 | 4.90 | 4.82 | 4.76 | 4.57 | 4.40 | 4.16 |
| 7 | 0.10 | 0.05 | 5.59 | 4.74 | 4.35 | 4.12 | 3.97 | 3.87 | 3.79 | 3.73 | 3.68 | 3.64 | 3.51 | 3.40 | 3.24 |
| 8 | 0.05 | 0.25 | 7.57 | 6.06 | 5.42 | 5.05 | 4.82 | 4.65 | 4.53 | 4.43 | 4.36 | 4.30 | 4.10 | 3.94 | 3.68 |
| 8 | 0.10 | 0.05 | 5.32 | 4.46 | 4.07 | 3.84 | 3.69 | 3.58 | 3.50 | 3.44 | 3.39 | 3.35 | 3.22 | 3.11 | 2.94 |
| 9 | 0.05 | 0.025 | 7.21 | 5.71 | 5.08 | 4.72 | 4.48 | 4.32 | 4.20 | 4.10 | 4.03 | 3.96 | 3.77 | 3.60 | 3.35 |
| 9 | 0.10 | 0.05 | 5.12 | 4.26 | 3.86 | 3.63 | 3.48 | 3.37 | 3.29 | 3.23 | 3.18 | 3.14 | 3.01 | 2.89 | 2.72 |
| 10 | 0.05 | 0.025 | 6.94 | 5.46 | 4.83 | 4.47 | 4.24 | 4.07 | 3.95 | 3.85 | 3.78 | 3.72 | 3.52 | 3.35 | 3.09 |
| 10 | 0.10 | 0.05 | 4.96 | 4.10 | 3.71 | 3.48 | 3.33 | 3.22 | 3.14 | 3.07 | 3.02 | 2.98 | 2.85 | 2.73 | 2.55 |
| 15 | 0.05 | 0.025 | 6.20 | 4.77 | 4.15 | 3.80 | 3.58 | 3.41 | 3.29 | 3.20 | 3.12 | 3.06 | 2.86 | 2.69 | 2.41 |
| 15 | 0.10 | 0.05 | 4.54 | 3.68 | 3.29 | 3.06 | 2.90 | 2.79 | 2.71 | 2.64 | 2.59 | 2.54 | 2.40 | 2.28 | 2.08 |
| 20 | 0.05 | 0.025 | 5.87 | 4.46 | 3.86 | 3.51 | 3.29 | 3.13 | 3.01 | 2.91 | 2.84 | 2.77 | 2.57 | 2.40 | 2.10 |
| 20 | 0.10 | 0.05 | 4.35 | 3.49 | 3.10 | 2.87 | 2.71 | 2.60 | 2.51 | 2.45 | 2.39 | 2.35 | 2.20 | 2.07 | 1.86 |
| 30 | 0.05 | 0.025 | 5.57 | 4.18 | 3.59 | 3.25 | 3.03 | 2.87 | 2.75 | 2.65 | 2.57 | 2.51 | 2.31 | 2.12 | 1.81 |

(continued)

| *df* of denominator | 2-tailed | L-tailed | Degrees of freedom (*df*) of the numerator | | | | | | | | | | | | |
|---|---|---|---|---|---|---|---|---|---|---|---|---|---|---|---|
| | *P*-value | *P*-value | 1 | 2 | 3 | 4 | 5 | 6 | 7 | 8 | 9 | 10 | 15 | 25 | 500 |
| 30 | 0.10 | 0.05 | 4.17 | 3.32 | 2.92 | 2.69 | 2.53 | 2.42 | 2.33 | 2.27 | 2.21 | 2.16 | 2.01 | 1.88 | 1.64 |
| 50 | 0.05 | 0.025 | 5.34 | 3.97 | 3.39 | 3.05 | 2.83 | 2.67 | 2.55 | 2.46 | 2.38 | 2.32 | 2.11 | 1.92 | 1.57 |
| 50 | 0.10 | 0.05 | 4.03 | 3.18 | 2.79 | 2.56 | 2.40 | 2.29 | 2.20 | 2.13 | 2.07 | 2.03 | 1.87 | 1.73 | 1.46 |
| 100 | 0.05 | 0.025 | 5.18 | 3.83 | 3.25 | 2.92 | 2.70 | 2.54 | 2.42 | 2.32 | 2.24 | 2.18 | 1.97 | 1.77 | 1.38 |
| 100 | 0.10 | 0.05 | 3.94 | 3.09 | 2.70 | 2.46 | 2.31 | 2.19 | 2.10 | 2.03 | 1.97 | 1.93 | 1.77 | 1.62 | 1.31 |
| 100 | 0.10 | 0.05 | 3.94 | 3.09 | 2.70 | 2.46 | 2.31 | 2.19 | 2.10 | 1.03 | 1.97 | 1.93 | 1.77 | 1.62 | 1.31 |
| 1000 | 0.05 | 0.025 | 5.04 | 3.70 | 3.13 | 2.80 | 2.58 | 2.42 | 2.30 | 2.20 | 2.13 | 2.06 | 1.85 | 1.64 | 1.16 |
| 1000 | 0.10 | 0.05 | 3.85 | 3.00 | 2.16 | 2.38 | 2.22 | 2.11 | 2.02 | 1.95 | 1.89 | 1.84 | 1.68 | 1.52 | 1.13 |

## 6 Conclusion

In some clinical studies, the spread of the data may be more relevant than the average of the data. For example, the spread of glucose levels of a slow-release-insulin is important. This chapter assesses how the spread of one and two data-samples can be estimated. In the Chap. 22 statistical tests for variability assessments with three or more samples will be given.

## 7 Note

More background, theoretical and mathematical information of variability assessments is given in Statistics applied to clinical studies 5th edition, Chap. 44, Springer Heidelberg Germany, 2012, from the same authors.

# Chapter 22
# Variability Analysis for Three or More Samples

## 1 General Purpose

In some clinical studies, the spread of the data may be more relevant than the average of the data. E.g., when we assess how a drug reaches various organs, variability of drug concentrations is important, as in some cases too little and in other cases dangerously high levels get through. Also, variabilities in drug response may be important. For example, the spread of glucose levels of a slow-release-insulin is important. In Chap. 21, the chi-square test for one sample and the F-test for two samples have been explained. In this chapter we will explain the Bartlett's test which is suitable for comparing three or more samples.

## 2 Schematic Overview of Type of Data File

| Outcome | Treatment group (1, 2, 3, ....) |
|---|---|
| . | . |
| . | . |
| . | . |
| . | . |
| . | . |
| . | . |
| . | . |
| . | . |
| . | . |
| | |
| | |
| | |
| | |
| | |



T.J. Cleophas, A.H. Zwinderman, *Clinical Data Analysis on a Pocket Calculator*,
DOI 10.1007/978-3-319-27104-0_22

## 3 Primary Scientific Question

Is the difference in variabilities between three or more samples statistically significant.

## 4 Data Example (Bartlett's Test)

The Bartlett's test, appropriate for comparing multiple samples for differences in variabilities, uses the underneath equation ($\chi^2 =$ chi-square value).

$$\chi^2 = (n_1 + n_2 + n_3 - 3)\ln s^2 - \left[(n_1 - 1)\ln s_1^2 + (n_2 - 1)\ln s_2^2 + (n_3 - 1)\ln s_3^2\right]$$

where

$n_1 =$ size sample 1
$s_1^2 =$ variance sample 1
$s^2 =$ pooled variance $= \frac{(n_1-1)s_1^2+(n_2-1)s_2^2+(n_3-1)s_3^2}{n_1+n_2+n_3-3} =$
$\ln =$ natural logarithm

As an example, blood glucose variabilities are assessed in a parallel-group study of three insulin treatment regimens. For that purpose three different groups of patients are treated with different insulin regimens. Variabilities of blood glucose levels are estimated by group-variances ($\ln =$ natural logarithm):

| | Group size (n) | Variance [(mmol/l)$^2$] |
|---|---|---|
| Group 1 | 100 | 8.0 |
| Group 2 | 100 | 14.0 |
| Group 3 | 100 | 18.0 |

$$\text{Pooled variance} = \frac{99 \times 8.0 + 99 \times 14.0 + 99 \times 18.0}{297} = 13.333$$

$$\begin{aligned}\chi^2 &= 297 \times \ln 13.333 - 99 \times \ln 8.0 - 99 \times \ln 14.0 - 99 \times \ln 18.0 = \\ &297 \times 2.58776 - 99 \times 2.079 - 99 \times 2.639 - 99 \times 2.890 = \\ &768.58 - 753.19 = \\ &15.37\end{aligned}$$

We have three separate groups, and, so, 3–1 = 2 degrees of freedom. The underneath chi-square table has an upper row with areas under the curve, a left-end column with degrees of freedom, and a whole lot of chi-square values. It shows that with a chi-square value of 15.37 a very significant difference between the three variances is demonstrated at $p < 0.001$. If the three groups are representative comparable samples, then we may conclude, that these three insulin regimens do not produce the same spread of glucose levels.

Chi-squared distribution

| Two-tailed *P*-value | | | | |
|---|---|---|---|---|
| *df* | 0.10 | 0.05 | 0.01 | 0.001 |
| 1 | 2.706 | 3.841 | 6.635 | 10.827 |
| 2 | 4.605 | 5.991 | 9.210 | 13.815 |
| 3 | 6.251 | 7.851 | 11.345 | 16.266 |
| 4 | 7.779 | 9.488 | 13.277 | 18.466 |
| 5 | 9.236 | 11.070 | 15.086 | 20.515 |
| 6 | 10.645 | 12.592 | 16.812 | 22.457 |
| 7 | 12.017 | 14.067 | 18.475 | 24.321 |
| 8 | 13.362 | 15.507 | 20.090 | 26.124 |
| 9 | 14.684 | 16.919 | 21.666 | 27.877 |
| 10 | 15.987 | 18.307 | 23.209 | 29.588 |
| 11 | 17.275 | 19.675 | 24.725 | 31.264 |
| 12 | 18.549 | 21.026 | 26.217 | 32.909 |
| 13 | 19.812 | 22.362 | 27.688 | 34.527 |
| 14 | 21.064 | 23.685 | 29.141 | 36.124 |
| 15 | 22.307 | 24.996 | 30.578 | 37.698 |
| 16 | 23.542 | 26.296 | 32.000 | 39.252 |
| 17 | 24.769 | 27.587 | 33.409 | 40.791 |
| 18 | 25.989 | 28.869 | 34.805 | 42.312 |
| 19 | 27.204 | 30.144 | 36.191 | 43.819 |
| 20 | 28.412 | 31.410 | 37.566 | 45.314 |
| 21 | 29.615 | 32.671 | 38.932 | 46.796 |
| 22 | 30.813 | 33.924 | 40.289 | 48.268 |
| 23 | 32.007 | 35.172 | 41.638 | 49.728 |
| 24 | 33.196 | 36.415 | 42.980 | 51.179 |
| 25 | 34.382 | 37.652 | 44.314 | 52.619 |
| 26 | 35.536 | 38.885 | 45.642 | 54.051 |
| 27 | 36.741 | 40.113 | 46.963 | 55.475 |
| 28 | 37.916 | 41.337 | 48.278 | 56.892 |
| 29 | 39.087 | 42.557 | 49.588 | 58.301 |
| 30 | 40.256 | 43.773 | 50.892 | 59.702 |
| 40 | 51.805 | 55.758 | 63.691 | 73.403 |
| 50 | 63.167 | 67.505 | 76.154 | 86.660 |
| 60 | 74.397 | 79.082 | 88.379 | 99.608 |
| 70 | 85.527 | 90.531 | 100.43 | 112.32 |
| 80 | 96.578 | 101.88 | 112.33 | 124.84 |
| 90 | 107.57 | 113.15 | 124.12 | 137.21 |
| 100 | 118.50 | 124.34 | 135.81 | 149.45 |

## 5 Conclusion

An alternative to the Bartlett's test is the Levene's test. The Levene's test is less sensitive than the Bartlett's test to departures from normality. If there is a strong evidence that the data do in fact come from a normal, or nearly normal, distribution, then Bartlett's test has a better performance. Levene's test requires a lot of arithmetic, and is usually performed using statistical software. E.g., it is routinely used by SPSS when performing an unpaired t-test or one-way ANOVA (analysis of variance) (see also Cleophas, Zwinderman, SPSS for Starters, part 1, Springer New York, 2010, Chaps. 4 and 8).

We should add that assessing significance of differences between 3 or more variances does not answer which of the samples produced the best outcome. Just like with analysis of variance (Chap. 19), separate post hoc one by one analyses are required.

## 6 Note

More background, theoretical and mathematical information of variability assessments is given in Statistics applied to clinical studies 5th edition, Chap. 44, Springer Heidelberg Germany, 2012, from the same authors.

# Chapter 23
# Confounding

## 1 General Purpose

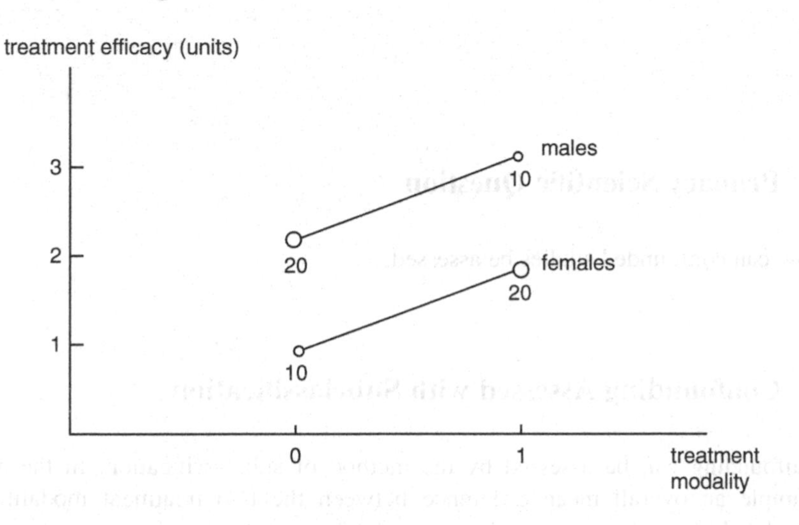


In the above study the treatment effects are better in the males than they are in the females. This difference in efficacy does not influence the overall assessment as long as the numbers of males and females in the treatment comparison are equally distributed. If, however, many females received the new treatment, and many males received the control treatment, a peculiar effect on the overall data analysis is observed as demonstrated by the difference in magnitudes of the circles in the above figure: the overall regression line will become close to horizontal, giving rise to the erroneous conclusion that no difference in efficacy exists between treatment and control. This phenomenon is called confounding, and may have a profound effect on the outcome of the study. This chapter shows how to assess confounded studies with continuous outcome data. Confounded studies with binary outcome data are reviewed in the Chap. 40.



T.J. Cleophas, A.H. Zwinderman, *Clinical Data Analysis on a Pocket Calculator*,
DOI 10.1007/978-3-319-27104-0_23

## 2 Schematic Overview of Type of Data File

| Outcome | Predictor (treatment modality) |
|---|---|
| . | . |
| . | . |
| . | . |
| . | . |
| . | . |
| . | . |
| . | . |
| . | . |
| . | . |
| | |
| | |
| | |
| | |
| | |

## 3 Primary Scientific Question

How can confounded studies be assessed.

## 4 Confounding Assessed with Subclassification

Confounding can be assessed by the method of subclassification. In the above example an overall mean difference between the two treatment modalities is calculated.

For treatment zero

Mean effect ± standard error (SE) = 1.5 units ± 0.5 units

For treatment one

Mean effect ± SE = 2.5 units ± 0.6 units

The mean difference of the two treatments

$$= 1.0 \text{ units} \pm \text{pooled standard error}$$
$$= 1.0 \pm \sqrt{(0.5^2 + 0.6^2)}$$
$$= 1.0 \pm 0.61$$

The t-value as calculated $= 1.0 / 0.61 = 1.639$

With 100-2 (100 patients, 2 groups) = 98 degrees of freedom the p-value of difference is calculated.

| df | One-Tail = .4 | .25 | .1 | .05 | .025 | .01 | .005 | .0025 | .001 | .0005 |
|---|---|---|---|---|---|---|---|---|---|---|
| | Two-Tail = .8 | .5 | .2 | .1 | .05 | .02 | .01 | .005 | .002 | .001 |
| 1 | 0.325 | 1.000 | 3.078 | 6.314 | 12.706 | 31.821 | 63.657 | 127.32 | 318.31 | 636.62 |
| 2 | 0.289 | 0.816 | 1.886 | 2.920 | 4.303 | 6.965 | 9.925 | 14.089 | 22.327 | 31.598 |
| 3 | 0.277 | 0.765 | 1.638 | 2.353 | 3.182 | 4.541 | 5.841 | 7.453 | 10.214 | 12.924 |
| 4 | 0.271 | 0.741 | 1.533 | 2.132 | 2.776 | 3.747 | 4.604 | 5.598 | 7.173 | 8.610 |
| 5 | 0.267 | 0.727 | 1.476 | 2.015 | 2.571 | 3.365 | 4.032 | 4.773 | 5.893 | 6.869 |
| 6 | 0.265 | 0.718 | 1.440 | 1.943 | 2.447 | 3.143 | 3.707 | 4.317 | 5.208 | 5.959 |
| 7 | 0.263 | 0.711 | 1.415 | 1.895 | 2.365 | 2.998 | 3.499 | 4.029 | 4.785 | 5.408 |
| 8 | 0.262 | 0.706 | 1.397 | 1.860 | 2.306 | 2.896 | 3.355 | 3.833 | 4.501 | 5.041 |
| 9 | 0.261 | 0.703 | 1.383 | 1.833 | 2.262 | 2.821 | 3.250 | 3.690 | 4.297 | 4.781 |
| 10 | 0.260 | 0.700 | 1.372 | 1.812 | 2.228 | 2.764 | 3.169 | 3.581 | 4.144 | 4.587 |
| 11 | 0.260 | 0.697 | 1.363 | 1.796 | 2.201 | 2.718 | 3.106 | 3.497 | 4.025 | 4.437 |
| 12 | 0.259 | 0.695 | 1.356 | 1.782 | 2.179 | 2.681 | 3.055 | 3.428 | 3.930 | 4.318 |
| 13 | 0.259 | 0.694 | 1.350 | 1.771 | 2.160 | 2.650 | 3.012 | 3.372 | 3.852 | 4.221 |
| 14 | 0.258 | 0.692 | 1.345 | 1.761 | 2.145 | 2.624 | 2.977 | 3.326 | 3.787 | 4.140 |
| 15 | 0.258 | 0.691 | 1.341 | 1.753 | 2.131 | 2.602 | 2.947 | 3.286 | 3.733 | 4.073 |
| 16 | 0.258 | 0.690 | 1.337 | 1.746 | 2.120 | 2.583 | 2.921 | 3.252 | 3.686 | 4.015 |
| 17 | 0.257 | 0.689 | 1.333 | 1.740 | 2.110 | 2.567 | 2.898 | 3.222 | 3.646 | 3.965 |
| 18 | 0.257 | 0.688 | 1.330 | 1.734 | 2.101 | 2.552 | 2.878 | 3.197 | 3.610 | 3.922 |
| 19 | 0.257 | 0.688 | 1.328 | 1.729 | 2.093 | 2.539 | 2.861 | 3.174 | 3.579 | 3.883 |
| 20 | 0.257 | 0.687 | 1.325 | 1.725 | 2.086 | 2.528 | 2.845 | 3.153 | 3.552 | 3.850 |
| 21 | 0.257 | 0.686 | 1.323 | 1.721 | 2.080 | 2.518 | 2.831 | 3.135 | 3.527 | 3.819 |
| 22 | 0.256 | 0.686 | 1.321 | 1.717 | 2.074 | 2.508 | 2.819 | 3.119 | 3.505 | 3.792 |
| 23 | 0.256 | 0.685 | 1.319 | 1.714 | 2.069 | 2.500 | 2.807 | 3.104 | 3.485 | 3.767 |
| 24 | 0.256 | 0.685 | 1.318 | 1.711 | 2.064 | 2.492 | 2.797 | 3.091 | 3.467 | 3.745 |
| 25 | 0.256 | 0.684 | 1.316 | 1.708 | 2.060 | 2.485 | 2.787 | 3.078 | 3.450 | 3.725 |
| 26 | 0.256 | 0.684 | 1.315 | 1.706 | 2.056 | 2.479 | 2.779 | 3.067 | 3.435 | 3.707 |
| 27 | 0.256 | 0.684 | 1.314 | 1.703 | 2.052 | 2.473 | 2.771 | 3.057 | 3.421 | 3.690 |
| 28 | 0.256 | 0.683 | 1.313 | 1.701 | 2.048 | 2.467 | 2.763 | 3.047 | 3.408 | 3.674 |
| 29 | 0.256 | 0.683 | 1.311 | 1.699 | 2.045 | 2.462 | 2.756 | 3.038 | 3.396 | 3.659 |
| 30 | 0.256 | 0.683 | 1.310 | 1.697 | 2.042 | 2.457 | 2.750 | 3.030 | 3.385 | 3.646 |
| 40 | 0.255 | 0.681 | 1.303 | 1.684 | 2.021 | 2.423 | 2.704 | 2.971 | 3.307 | 3.551 |
| 60 | 0.254 | 0.679 | 1.296 | 1.671 | 2.000 | 2.390 | 2.660 | 2.915 | 3.232 | 3.460 |
| 120 | 0.254 | 0.677 | 1.289 | 1.658 | 1.980 | 2.358 | 2.617 | 2.860 | 3.160 | 3.373 |
| ∞ | 0.253 | 0.674 | 1.282 | 1.645 | 1.960 | 2.326 | 2.576 | 2.807 | 3.090 | 3.291 |

The above t-table has a left-end column giving degrees of freedom (≈ sample sizes), and two top rows with p-values (areas under the curve = p - values), one-tail meaning that only one end of the curve, two-tail meaning that both ends are assessed simultaneously. The t-table is, furthermore, full of t-values, that, with ∞ degrees of freedom, are equal to z-values (Chap. 36). The t-values are to be understood as mean results of studies, but not expressed in mmol/l, kilograms, but in so-called SEM-units (Standard error of the mean units), that are obtained by dividing your mean result by its own standard error. With many degrees of freedom (large samples) the curve will be a little bit narrower, and more in agreement with nature.

We have 98 degrees of freedom, and, so, we will need the rows between 60 and 120 to estimate the area under curve of a t-value of 1.639. A t-value larger than 1.980 is required for an area under curve significantly different from zero ($=$ p-value $< 0.05$). Our area under the curve of 1.639 is much larger than 0.05. We cannot reject the null-hypothesis of no difference in the data here.

In order to assess the possibility of confounding, a weighted mean has to be calculated. The underneath equation is adequate for the purpose.

$$\text{Weighted mean} = \frac{\text{Difference}_{\text{males}}/\text{its SE}^2 + \text{Difference}_{\text{females}}/\text{its SE}^2}{1/\text{SE}^2{}_{\text{males}} + 1/\text{SE}^2{}_{\text{females}}}$$

For the males we find means of 2.0 and 3.0 units, for the females 1.0 and 2.0 units. The mean difference for the males and females separately are 1.0 and 1.0 as expected from the above figure. However, the pooled standard errors are different, for the males 0.4, and for the females 0.3 units.

According to the above equation a weighted t-value is calculated

$$\begin{aligned}
\text{Weighted mean} &= \frac{\left(1.0/0.4^2 + 1.0/0.3^2\right)}{\left(1/0.4^2 + 1/0.3^2\right)} \\
&= 1.0 \\
\text{Weighted SE}^2 &= 1/\left(1/0.4^2 + 1/0.3^2\right) \\
&= 0.0576 \\
\text{Weighted SE} &= 0.24 \\
\text{t-value} &= 1.0/0.24 = 4.16
\end{aligned}$$

According to the above t-table with 98 degrees of freedom a t-value of 4.16 is much larger than the critical t-value producing a $p < 0.05$. Our result is very significant: p-value $= < 0.001$.

The weighted mean is equal to the unweighted mean. However, its SE is much smaller. It means that after adjustment for confounding a very significant difference is observed.

## 5 Conclusion

Other methods for assessing confounding include multiple regression analysis and propensity score assessments. Particularly, with more than a single confounder these two methods are unavoidable. Propensity score assessments is covered in the Chap. 32.

# 6 Note

More background, theoretical and mathematical information of confounding is given in Statistics applied to clinical studies 5th edition, Chap. 28, Springer Heidelberg Germany, 2012, from the same authors.

# Chapter 24
# Propensity Scores and Propensity Score Matching for Assessing Multiple Confounders

## 1 General Purpose

In the Chap. 23 methods for assessing confounders were reviewed. Propensity score are ideal for assessing confounding, particularly, if multiple confounders are in a study. E.g., age and cardiovascular risk factors may not be similarly distributed in two treatment groups of a parallel-group study. Propensity score matching is used to make observational data look like randomized controlled trial data. This chapter assesses propensity score and propernsity score matching.

## 2 Schematic Overview of Type of Data File

| Outcome | Treatment modality | Propensity scores |
|---|---|---|
| . | . | . |
| . | . | . |
| . | . | . |
| . | . | . |
| . | . | . |
| . | . | . |
| . | . | . |
| . | . | . |
| . | . | . |
| | | |
| | | |
| | | |
| | | |
| | | |



T.J. Cleophas, A.H. Zwinderman, *Clinical Data Analysis on a Pocket Calculator*,
DOI 10.1007/978-3-319-27104-0_24

## 3 Primary Scientific Question

Is propensity score and propensity score matching adequate for assessing studies with multiple confounders.

## 4 Propensity Scores

A propensity (prop) score for age can be defined as the risk ratio (or rather odds ratio) of receiving treatment 1 compared to that of treatment 2 if you are old in this study.

| | Treatment-1 | Treatment-2 | odds treatment-1 / odds treatment-2 |
|---|---|---|---|
| | n = 100 | n = 100 | (OR) |
| 1. Age > 65 | 63 | 76 | 0.54 (63/76 / 37/24) |
| 2. Age < 65 | 37 | 24 | 1.85 ( $= OR_2 = 1/OR_1$) |
| 3. Diabetes | 20 | 33 | 0.51 |
| 4. Not diabetes | 80 | 67 | 1.96 |
| 5. Smoker | 50 | 80 | 0.25 |
| 6. Not smoker | 50 | 20 | 4.00 |
| 7. Hypertension | 51 | 65 | 0.65 |
| 8. Not hypertension | 49 | 35 | 1.78 |
| 10. Not cholesterol | 39 | 22 | 2.27 |

The odds ratios can be tested for statistical significance (see Chap. 2, odds ratios), and those that are statistically significant can, then, be used for calculating a combined propensity-score for all of the inequal characteristics by multiplying the significant odds ratios, and, then, calculating from this product the combined propensity-score = combined "risk ratio" (= combined OR / (1+ combined OR). y = yes, n = no, combined OR = $OR_1$ x $OR_3$ x $OR_5$ x $OR_7$ x $OR_9$.

| | Old | Diab | Smoker | Hypert | Cholesterol | Combined OR | Combined propensity score |
|---|---|---|---|---|---|---|---|
| Patient 1 | y | y | n | y | y | 7.99 | 0.889 |
| 2 | n | n | n | y | y | 105.27 | 0.991 |
| 3 | y | n | n | y | y | 22.80 | 0.958 |
| 4 | y | y | y | y | y | 0.4999 | 0.333 |
| 5 | n | n | y | | | | |
| 6 | y | y | y | | | | |
| 7 | .... | | | | | | |
| 8 | .... | | | | | | |

Each patient has his / her own propensity score based on and adjusted for the significantly larger chance of receiving one treatment versus the other treatment.

Usually, propensity score adjustment for confounders is accomplished by dividing the patients into four subgroups, but for the purpose of simplicity we here use 2 subgroups, those with high and those with low propensity scores.

Confounding is assessed by the method of subclassification. In the above example an overall mean difference between the two treatment modalities is calculated.

For treatment zero

Mean effect $\pm$ standard error (SE) $= 1.5$ units $\pm\, 0.5$ units

For treatment one

Mean effect $\pm$ SE $= 2.5$ units $\pm\, 0.6$ units

The mean difference of the two treatments

$= 1.0$ units $\pm$ pooled standard error

$= 1.0 \pm \sqrt{(0.5^2 + 0.6^2)}$

$= 1.0 \pm 0.61$

The t-value as calculated $= 1.0/0.61 = 1.639$

The underneath t-table is helpful to determine a p-value.

| df | One-Tail = .4<br>Two-Tail = .8 | .25<br>.5 | .1<br>.2 | .05<br>.1 | .025<br>.05 | .01<br>.02 | .005<br>.01 | .0025<br>.005 | .001<br>.002 | .0005<br>.001 |
|---|---|---|---|---|---|---|---|---|---|---|
| 1 | 0.325 | 1.000 | 3.078 | 6.314 | 12.706 | 31.821 | 63.657 | 127.32 | 318.31 | 636.62 |
| 2 | 0.289 | 0.816 | 1.886 | 2.920 | 4.303 | 6.965 | 9.925 | 14.089 | 22.327 | 31.598 |
| 3 | 0.277 | 0.765 | 1.638 | 2.353 | 3.182 | 4.541 | 5.841 | 7.453 | 10.214 | 12.924 |
| 4 | 0.271 | 0.741 | 1.533 | 2.132 | 2.776 | 3.747 | 4.604 | 5.598 | 7.173 | 8.610 |
| 5 | 0.267 | 0.727 | 1.476 | 2.015 | 2.571 | 3.365 | 4.032 | 4.773 | 5.893 | 6.869 |
| 6 | 0.265 | 0.718 | 1.440 | 1.943 | 2.447 | 3.143 | 3.707 | 4.317 | 5.208 | 5.959 |
| 7 | 0.263 | 0.711 | 1.415 | 1.895 | 2.365 | 2.998 | 3.499 | 4.029 | 4.785 | 5.408 |
| 8 | 0.262 | 0.706 | 1.397 | 1.860 | 2.306 | 2.896 | 3.355 | 3.833 | 4.501 | 5.041 |
| 9 | 0.261 | 0.703 | 1.383 | 1.833 | 2.262 | 2.821 | 3.250 | 3.690 | 4.297 | 4.781 |
| 10 | 0.260 | 0.700 | 1.372 | 1.812 | 2.228 | 2.764 | 3.169 | 3.581 | 4.144 | 4.587 |
| 11 | 0.260 | 0.697 | 1.363 | 1.796 | 2.201 | 2.718 | 3.106 | 3.497 | 4.025 | 4.437 |
| 12 | 0.259 | 0.695 | 1.356 | 1.782 | 2.179 | 2.681 | 3.055 | 3.428 | 3.930 | 4.318 |
| 13 | 0.259 | 0.694 | 1.350 | 1.771 | 2.160 | 2.650 | 3.012 | 3.372 | 3.852 | 4.221 |
| 14 | 0.258 | 0.692 | 1.345 | 1.761 | 2.145 | 2.624 | 2.977 | 3.326 | 3.787 | 4.140 |
| 15 | 0.258 | 0.691 | 1.341 | 1.753 | 2.131 | 2.602 | 2.947 | 3.286 | 3.733 | 4.073 |
| 16 | 0.258 | 0.690 | 1.337 | 1.746 | 2.120 | 2.583 | 2.921 | 3.252 | 3.686 | 4.015 |
| 17 | 0.257 | 0.689 | 1.333 | 1.740 | 2.110 | 2.567 | 2.898 | 3.222 | 3.646 | 3.965 |
| 18 | 0.257 | 0.688 | 1.330 | 1.734 | 2.101 | 2.552 | 2.878 | 3.197 | 3.610 | 3.922 |
| 19 | 0.257 | 0.688 | 1.328 | 1.729 | 2.093 | 2.539 | 2.861 | 3.174 | 3.579 | 3.883 |
| 20 | 0.257 | 0.687 | 1.325 | 1.725 | 2.086 | 2.528 | 2.845 | 3.153 | 3.552 | 3.850 |
| 21 | 0.257 | 0.686 | 1.323 | 1.721 | 2.080 | 2.518 | 2.831 | 3.135 | 3.527 | 3.819 |
| 22 | 0.256 | 0.686 | 1.321 | 1.717 | 2.074 | 2.508 | 2.819 | 3.119 | 3.505 | 3.792 |
| 23 | 0.256 | 0.685 | 1.319 | 1.714 | 2.069 | 2.500 | 2.807 | 3.104 | 3.485 | 3.767 |
| 24 | 0.256 | 0.685 | 1.318 | 1.711 | 2.064 | 2.492 | 2.797 | 3.091 | 3.467 | 3.745 |
| 25 | 0.256 | 0.684 | 1.316 | 1.708 | 2.060 | 2.485 | 2.787 | 3.078 | 3.450 | 3.725 |
| 26 | 0.256 | 0.684 | 1.315 | 1.706 | 2.056 | 2.479 | 2.779 | 3.067 | 3.435 | 3.707 |
| 27 | 0.256 | 0.684 | 1.314 | 1.703 | 2.052 | 2.473 | 2.771 | 3.057 | 3.421 | 3.690 |
| 28 | 0.256 | 0.683 | 1.313 | 1.701 | 2.048 | 2.467 | 2.763 | 3.047 | 3.408 | 3.674 |
| 29 | 0.256 | 0.683 | 1.311 | 1.699 | 2.045 | 2.462 | 2.756 | 3.038 | 3.396 | 3.659 |
| 30 | 0.256 | 0.683 | 1.310 | 1.697 | 2.042 | 2.457 | 2.750 | 3.030 | 3.385 | 3.646 |
| 40 | 0.255 | 0.681 | 1.303 | 1.684 | 2.021 | 2.423 | 2.704 | 2.971 | 3.307 | 3.551 |
| 60 | 0.254 | 0.679 | 1.296 | 1.671 | 2.000 | 2.390 | 2.660 | 2.915 | 3.232 | 3.460 |
| 120 | 0.254 | 0.677 | 1.289 | 1.658 | 1.980 | 2.358 | 2.617 | 2.860 | 3.160 | 3.373 |
| ∞ | 0.253 | 0.674 | 1.282 | 1.645 | 1.960 | 2.326 | 2.576 | 2.807 | 3.090 | 3.291 |

The t-table has a left-end column giving degrees of freedom (≈ sample sizes), and two top rows with p-values (areas under the curve = p − values), one-tail meaning that only one end of the curve, two-tail meaning that both ends are assessed simulataneously. The t-table is, furthermore, full of t-values, that, with ∞ degrees of freedom, are equal to z-values (Chap. 36). The t-values are to be understood as mean results of studies, but not expressed in mmol/l, kilograms, but in so-called SEM-units (Standard error of the mean units), that are obtained by dividing your mean result by its own standard error. With many degrees of freedom (large samples) the curve will be a little bit narrower, and more in agreement with nature.

With 200–2 (200 patients, 2 groups) = 198 degrees of freedom, a t-value > 1.96 is required to obtain a two-sided $p < 0.05$. It can be observed that our p-value > 0.05. It is even > 0.10.

In order to assess the possibility of confounding, a weighted mean has to be calculated. The underneath equation is adequate for the purpose (prop score = propensity score).

$$\text{Weighted mean} = \frac{\text{Difference}_{\text{high prop score}} / \text{its SE}^2 + \text{Difference}_{\text{low prop score}} / \text{its SE}^2}{1/\text{SE}^2_{\text{high prop score}} + 1/\text{SE}^2_{\text{low prop score}}}$$

For the high prop score we find means of 2.0 and 3.0 units, for the low prop score 1.0 and 2.0 units. The mean difference separately are 1.0 and 1.0 as expected. However, the pooled standard errors are different, for the males 0.4, and for the females 0.3 units.

According to the above equation a weighted t-value is calculated

$$\begin{aligned}
\text{Weighted mean} &= \frac{(1.0/0.4^2 + 1.0/0.3^2)}{(1/0.4^2 + 1/0.3^2)} \\
&= 1.0 \\
\text{Weighted SE}^2 &= 1/(1/0.4^2 + 1/0.3^2) \\
&= 0.0576 \\
\text{Weighted SE} &= 0.24 \\
\text{t-value} &= 1.0/0.24 = 4.16
\end{aligned}$$

With 98 degrees of freedom, and a t-value of 4.16 means a two sided p-value < 0.001 is obtained.

The weighted mean is equal to the unweighted mean. However, its SE is much smaller. It means that after adjustment for the prop scores a very significant difference is observed. Instead of subclassification, also linear regression with the propensity scores as covariate is a common way to deal with propensity scores. However, this is hard on a pocket calculator.

## 5 Propensity Score Matching

In the study of 200 patients each patient has his/her own propensity score. We select for each patient in group 1 a patient from group 2 with the same propensity score.

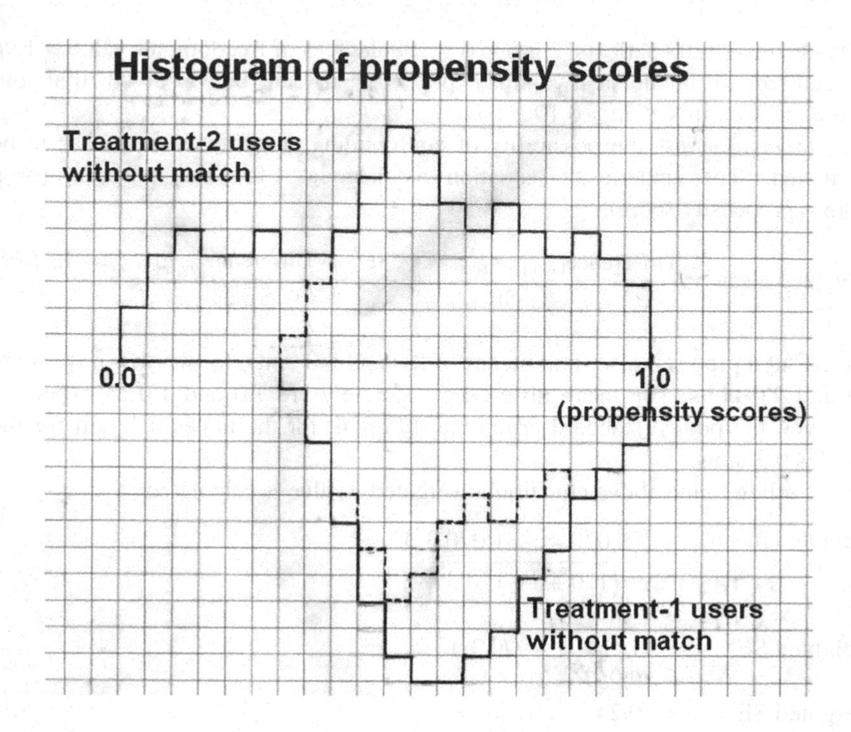


The above graph is an example of the nearest neighbor watching method for matching patients with similar propensity scores. Each square represents one patient. In random order the first patient from group 1 is selected. Then, he/she is matched to the patient of group 2 with the nearest propensity score. We will continue until there are no longer similar propensity scores. Group 1 has to be summarized above the x-axis, group 2 below it. The patients with dissimilar propensity scores that cannot be matched, have to be removed from the analysis.

This procedure will end up sampling two new groups that are entirely symmetric on their subgroup variables, and can, thus, be simply analyzed as two groups in a randomized trial. In the given example two matched groups of 71 patients were left for comparison of the treatments. They can be analyzed for treatment differences using unpaired t-tests (Chap. 7) or chi-square tests (Chap. 38), without the need to further account confounding anymore.

## 6 Conclusion

Propensity score are for assessing studies with multiple confounding variables, e.g., age and cardiovascular risk factors, factors that are likely not to be similarly distributed in two treatment groups of a parallel-group study. Propensity score matching is used to make observational data look like randomized controlled trial data. This chapter assesses propensity score and propernsity score matching.

## 7 Note

More background, theoretical and mathematical information of propensity scores is given in Statistics applied to clinical studies 5th edition, Chap. 29, Springer Heidelberg Germany, 2012, from the same authors.

# Chapter 25
# Interaction

## 1 General Purpose

The medical concept of interaction is synonymous to the terms heterogeneity and synergism. Interaction must be distinguished from confounding. In a trial with interaction effects the parallel groups have similar characteristics. However, there are subsets of patients that have an unusually high or low response. The figure below gives an example of a study in which males seem to respond better to the treatment 1 than females. With confounding things are different. For whatever reason the randomization has failed, the parallel groups have asymmetric characteristics. E.g., in a placebo-controlled trial of two parallel-groups asymmetry of age may be a confounder. The control group is significantly older than the treatment group, and this can easily explain the treatment difference as demonstrated in the previous chapter. This chapter uses simply t-tests for assessing interactions.



T.J. Cleophas, A.H. Zwinderman, *Clinical Data Analysis on a Pocket Calculator*,
DOI 10.1007/978-3-319-27104-0_25

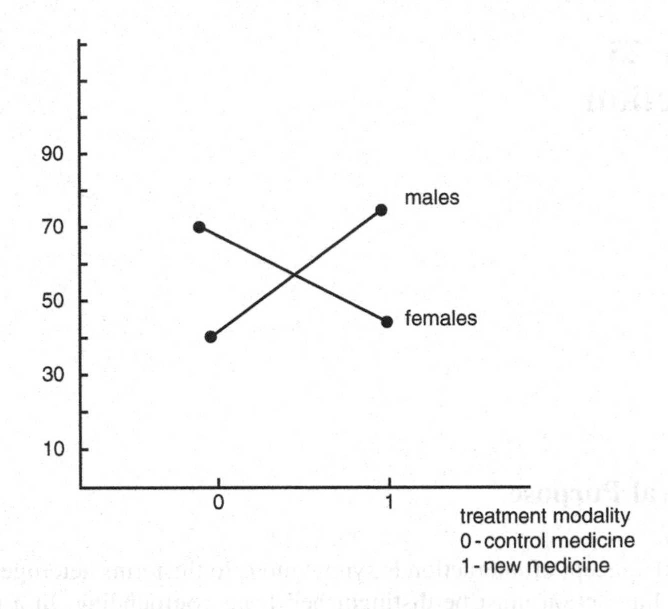


## 2 Schematic Overview of Type of Data File

| Outcome | Exposure 1 | Exposure 2 |
|---|---|---|
| . | . | . |
| . | . | . |
| . | . | . |
| . | . | . |
| . | . | . |
| . | . | . |
| . | . | . |
| . | . | . |
| . | . | . |
| | | |
| | | |

## 3 Primary Scientific Question

Can t-tests adequately demonstrate interaction between the effect on the outcome of two exposure variables.

# 4 Data Example of Interaction

A parallel-group study of verapamil versus metoprolol for the treatment of paroxysmal atrial tachycardias is given below. The numbers of episodes of paroxysmal atrial tachycardias per patient are the outcome variable.

| | VERAPAMIL | | METOPROLOL | | |
|---|---|---|---|---|---|
| MALES | 52 | | 28 | | |
| | 48 | | 35 | | |
| | 43 | | 34 | | |
| | 50 | | 32 | | |
| | 43 | | 34 | | |
| | 44 | | 27 | | |
| | 46 | | 31 | | |
| | 46 | | 27 | | |
| | 43 | | 29 | | |
| | 49 | | 25 | | |
| | | 464 | | 302 | 766 |
| FEMALES | 38 | | 43 | | |
| | 42 | | 34 | | |
| | 42 | | 33 | | |
| | 35 | | 42 | | |
| | 33 | | 41 | | |
| | 38 | | 37 | | |
| | 39 | | 37 | | |
| | 34 | | 40 | | |
| | 33 | | 36 | | |
| | 34 | | 35 | | |
| | | 368 | | 378 | 746 |
| | 832 | | 680 | | |

Overall metoprolol seems to perform better. However, this is only true only for one subgroup (males). SD = standard deviation, SE = standard error.

| | Males | Females |
|---|---|---|
| Mean $_{verapamil}$ (SD) | 46.4 (3.23866) | 36.8 (3.489667) |
| Mean $_{metoprolol}$ (SD) | 30.2 (3.48966)- | 37.8 (3.489667)- |
| Difference means (SE) | 16.2 (1.50554) | −1.0 (1.5606) |
| Difference between males and females | | 17.2 ($SE_1^2 + SE_1^2 = 2.166$) |
| | t-value = 17.2/2.166 = 8.... | |
| | $p < 0.0001$ | |

We conclude, that there is a significant difference between the males and females, and, thus, a significant interaction between gender and treat-efficacy.

| df | One-Tail = .4<br>Two-Tail = .8 | .25<br>.5 | .1<br>.2 | .05<br>.1 | .025<br>.05 | .01<br>.02 | .005<br>.01 | .0025<br>.005 | .001<br>.002 | .0005<br>.001 |
|---|---|---|---|---|---|---|---|---|---|---|
| 1 | 0.325 | 1.000 | 3.078 | 6.314 | 12.706 | 31.821 | 63.657 | 127.32 | 318.31 | 636.62 |
| 2 | 0.289 | 0.816 | 1.886 | 2.920 | 4.303 | 6.965 | 9.925 | 14.089 | 22.327 | 31.598 |
| 3 | 0.277 | 0.765 | 1.638 | 2.353 | 3.182 | 4.541 | 5.841 | 7.453 | 10.214 | 12.924 |
| 4 | 0.271 | 0.741 | 1.533 | 2.132 | 2.776 | 3.747 | 4.604 | 5.598 | 7.173 | 8.610 |
| 5 | 0.267 | 0.727 | 1.476 | 2.015 | 2.571 | 3.365 | 4.032 | 4.773 | 5.893 | 6.869 |
| 6 | 0.265 | 0.718 | 1.440 | 1.943 | 2.447 | 3.143 | 3.707 | 4.317 | 5.208 | 5.959 |
| 7 | 0.263 | 0.711 | 1.415 | 1.895 | 2.365 | 2.998 | 3.499 | 4.029 | 4.785 | 5.408 |
| 8 | 0.262 | 0.706 | 1.397 | 1.860 | 2.306 | 2.896 | 3.355 | 3.833 | 4.501 | 5.041 |
| 9 | 0.261 | 0.703 | 1.383 | 1.833 | 2.262 | 2.821 | 3.250 | 3.690 | 4.297 | 4.781 |
| 10 | 0.260 | 0.700 | 1.372 | 1.812 | 2.228 | 2.764 | 3.169 | 3.581 | 4.144 | 4.587 |
| 11 | 0.260 | 0.697 | 1.363 | 1.796 | 2.201 | 2.718 | 3.106 | 3.497 | 4.025 | 4.437 |
| 12 | 0.259 | 0.695 | 1.356 | 1.782 | 2.179 | 2.681 | 3.055 | 3.428 | 3.930 | 4.318 |
| 13 | 0.259 | 0.694 | 1.350 | 1.771 | 2.160 | 2.650 | 3.012 | 3.372 | 3.852 | 4.221 |
| 14 | 0.258 | 0.692 | 1.345 | 1.761 | 2.145 | 2.624 | 2.977 | 3.326 | 3.787 | 4.140 |
| 15 | 0.258 | 0.691 | 1.341 | 1.753 | 2.131 | 2.602 | 2.947 | 3.286 | 3.733 | 4.073 |
| 16 | 0.258 | 0.690 | 1.337 | 1.746 | 2.120 | 2.583 | 2.921 | 3.252 | 3.686 | 4.015 |
| 17 | 0.257 | 0.689 | 1.333 | 1.740 | 2.110 | 2.567 | 2.898 | 3.222 | 3.646 | 3.965 |
| 18 | 0.257 | 0.688 | 1.330 | 1.734 | 2.101 | 2.552 | 2.878 | 3.197 | 3.610 | 3.922 |
| 19 | 0.257 | 0.688 | 1.328 | 1.729 | 2.093 | 2.539 | 2.861 | 3.174 | 3.579 | 3.883 |
| 20 | 0.257 | 0.687 | 1.325 | 1.725 | 2.086 | 2.528 | 2.845 | 3.153 | 3.552 | 3.850 |
| 21 | 0.257 | 0.686 | 1.323 | 1.721 | 2.080 | 2.518 | 2.831 | 3.135 | 3.527 | 3.819 |
| 22 | 0.256 | 0.686 | 1.321 | 1.717 | 2.074 | 2.508 | 2.819 | 3.119 | 3.505 | 3.792 |
| 23 | 0.256 | 0.685 | 1.319 | 1.714 | 2.069 | 2.500 | 2.807 | 3.104 | 3.485 | 3.767 |
| 24 | 0.256 | 0.685 | 1.318 | 1.711 | 2.064 | 2.492 | 2.797 | 3.091 | 3.467 | 3.745 |
| 25 | 0.256 | 0.684 | 1.316 | 1.708 | 2.060 | 2.485 | 2.787 | 3.078 | 3.450 | 3.725 |
| 26 | 0.256 | 0.684 | 1.315 | 1.706 | 2.056 | 2.479 | 2.779 | 3.067 | 3.435 | 3.707 |
| 27 | 0.256 | 0.684 | 1.314 | 1.703 | 2.052 | 2.473 | 2.771 | 3.057 | 3.421 | 3.690 |
| 28 | 0.256 | 0.683 | 1.313 | 1.701 | 2.048 | 2.467 | 2.763 | 3.047 | 3.408 | 3.674 |
| 29 | 0.256 | 0.683 | 1.311 | 1.699 | 2.045 | 2.462 | 2.756 | 3.038 | 3.396 | 3.659 |
| 30 | 0.256 | 0.683 | 1.310 | 1.697 | 2.042 | 2.457 | 2.750 | 3.030 | 3.385 | 3.646 |
| 40 | 0.255 | 0.681 | 1.303 | 1.684 | 2.021 | 2.423 | 2.704 | 2.971 | 3.307 | 3.551 |
| 60 | 0.254 | 0.679 | 1.296 | 1.671 | 2.000 | 2.390 | 2.660 | 2.915 | 3.232 | 3.460 |
| 120 | 0.254 | 0.677 | 1.289 | 1.658 | 1.980 | 2.358 | 2.617 | 2.860 | 3.160 | 3.373 |
| ∞ | 0.253 | 0.674 | 1.282 | 1.645 | 1.960 | 2.326 | 2.576 | 2.807 | 3.090 | 3.291 |

The t-table has a left-end column giving degrees of freedom (≈ sample sizes), and two top rows with p-values (areas under the curve = p - values), one-tail meaning that only one end of the curve, two-tail meaning that both ends are assessed simulataneously. The t-table is, furthermore, full of t-values, that, with ∞ degrees of freedom, are equal to z-values (Chap. 36). The t-values are to be understood as mean results of studies, but not expressed in mmol/l, kilograms, but in so-called SEM-units (Standard error of the mean units), that are obtained by dividing your mean result by its own standard error. With many degrees of freedom (large samples) the curve will be a little bit narrower, and more in agreement with nature.

With 40–2 = 38 degrees of freedom (close to 40), and a t-value = 17.2/2.166 = 8.… and thus larger than 3.551, the two-tail p -value is $< 0.001$.

## 5 Conclusion

T-tests can readily be applied for assessing interactions between the effects on the outcome of two exposure variables. Interaction can also be assessed with analysis of variance and regression modeling. These two methods are the methods of choice in case you expect more than a single interaction in your data. They should be carried out on a computer.

## 6 Note

More background, theoretical and mathematical information of interaction assessments is given in Statistics applied to clinical studies 5th edition, Chap. 30, Springer Heidelberg Germany, 2012, from the same authors.

# Chapter 26
# Accuracy and Reliability Assessments

## 1 General Purpose

Clinical research is impossible without valid diagnostic tests. The methods for validating *qualitative* diagnostic tests, having binary outcomes, include sensitivity / specificity assessments and ROC (receiver operated characteristic) curves (Chaps. 52 and 53). In contrast, the methods for validating *quantitative* diagnostic tests, having continuous outcomes, have not been agreed upon. This chapter assesses pocket calculator methods for the purpose.

## 2 Schematic Overview of Type of Data File

| Gold standard test | New test |
|---|---|
| . | . |
| . | . |
| . | . |
| . | . |
| . | . |
| . | . |
| . | . |
| . | . |
| . | . |
| | |
| | |
| | |



T.J. Cleophas, A.H. Zwinderman, *Clinical Data Analysis on a Pocket Calculator*,
DOI 10.1007/978-3-319-27104-0_26

## 3 Primary Scientific Question

Is a new diagnostic test adequately accurate and adequately reliable.

## 4 Testing Accuracy with $R^2$-Values

Linear regression is often used for that purpose. The underneath figure gives an example. The regression equation is given by $y = a + b\ x$ ($a =$ intercept, $b =$ regression coefficient). More information of linear regression is given in the Chap. 8. In the underneath example given, the x-axis-data, ultrasound estimates, are a very significant predictor of the y-axis-data, the electromagnetic measurements. However, the prediction, despite the high level of statistical significance, is very imprecise. E.g., if $x = 6$, then y may be 10 or 21, and, if $x = 7$, then y may be 19, 31 or 32.

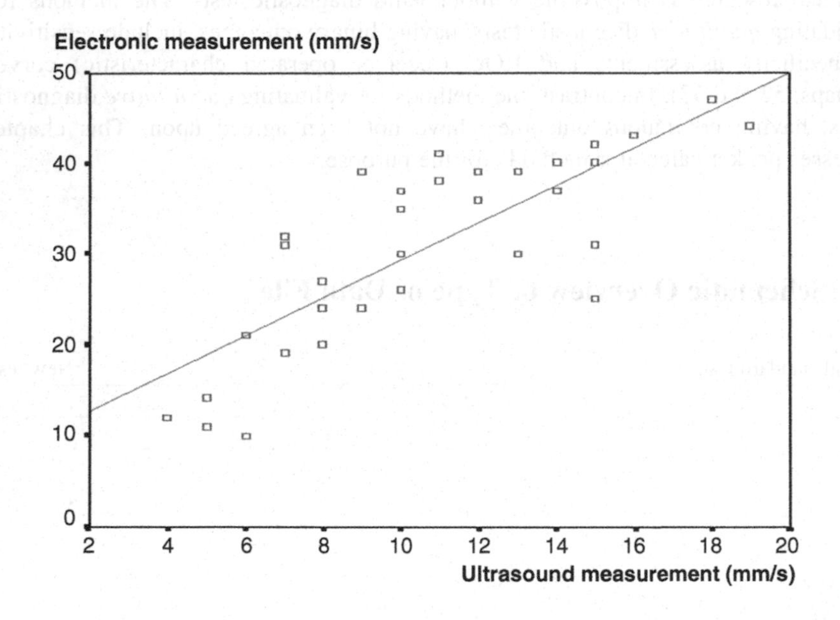


Probably, the best accuracy assessment is to test whether the 95 % confidence interval of the a-value (the intercept) is significantly different from 0, and whether the 95 % confidence interval of the b-value (the regression coefficient is significantly different from 1. However, second best and more easy on a pocket calculator is to test the data for a squared correlation coefficient ($r^2$) > 95 %. This method assumes a diagnostic test with the best fit equation $y = a + b\ x$, rather than $y = x$. A diagnostic test with the former best fit equation, like in the above example, is not necessarily useless, and could be approved as a valid test, if it is precise, that means, if the x-data precisely predict the *(y−a)/b- data* rather than the y-data. If we apply such a test, then the result of the x-data will, of course, have to be transformed into $a + b\,x$ in order to find the y-data.

We will use Electronic Calculator (see Chap. 1) for computations. First, we will calculate the b and r values.

Command:
click ON....click MODE....press 3....press 1....press SHIFT, MODE, and again 1.... press = ....start entering the data.... [x-datum$_1$ y-datum$_1$]....[x-datum$_2$ y-datum$_2$]....[..........] etc....

In order to obtain the r value, press: shift, S-VAR, ▶, ▶, 3, = .
The r-value equals 0.6...
The $r^2$ -value equals 0.36...

This $r^2$-value is much smaller than 0.95. It means that this diagnostic test can not be validated as being adequately accurate.

## 5 Testing Reliability with Duplicate Standard Deviations

The reliability, otherwise called reproducibility, of diagnostic tests is another important quality criterion. A diagnostic test is very unreliable, if it is not well-reproducible. Underneath a first data example is given.

| test 1 result | test 2 | difference | (difference)$^2$ |
|---|---|---|---|
| 1 | 11 | −10 | 100 |
| 10 | 0 | 10 | 100 |
| 2 | 11 | −9 | 81 |
| 12 | 2 | 10 | 100 |
| 11 | 1 | 10 | 100 |
| 1 | 12 | −11 | 121 |
| mean | | | |
| 6.17 | 6.17 | 0 | 100.3 |

Duplicate standard deviation $=\sqrt{(1/2 \times \text{mean of (difference)}^2)}$
$=\sqrt{(1/2 \times 100.3)}$
$=7.08$

The proportional duplicate standard deviation

$$= \frac{\text{duplicate standard deviation}}{\text{overall mean}} \times 100\,\%$$

$$= \frac{7.08}{6.17} \times 100\,\%$$
$$=115\,\%$$

An adequate reliability is obtained with a proportional duplicate standard deviation of 10–20 %. In the current example, although the mean difference between the two tests equals zero, there is, thus, a very poor reproducibility.

Underneath a second example is given. The question is, is this test well reproducible?

| test 1 result | test 2 |
|---|---|
| 6.2 | 5.1 |
| 7.0 | 7.8 |
| 8.1 | 3.9 |
| 7.5 | 5.5 |
| 6.5 | 6.6 |

Analysis:

| Test 1 Result | Test 2 | Difference | Difference$^2$ |
|---|---|---|---|
| 6.2 | 5.1 | 1.1 | 1.21 |
| 7.0 | 7.8 | −0.8 | 0.64 |
| 8.1 | 3.9 | 4.2 | 17.64 |
| 7.5 | 5.5 | 2.0 | 4.0 |
| 6.5 | 6.6 | −0.1 | 0.01 |
| Mean | | | |
| 7.06 | 5.78 | | 4.7 |
| grand mean 6.42 | | | |

Duplicate standard deviation $= \sqrt{(½ \times 4.7)}$
$= 1.553$

Proportional duplicate standard deviation %

$$= \frac{\text{duplicate standard deviation} \times 100\,\%}{\text{overall mean}}$$

$$= \frac{1.553}{6.42} \times 100\,\%$$

$$= 24\,\%$$

A good reproducibility is between 10 and 20 %. In the above example reproducibility is, thus, almost good.

## 6 Conclusion

In the current chapter two methods for validation of diagnostic methods with continuous data easily performed with the help of a pocket calculator are described. Many more methods exists. For accuracy assessments paired t-tests, Bland-Altman plots and the complex linear regession models of Passing-Bablok and Deming are available. For reliability repeatability coefficients and intraclass correlations are possible (see underneath "Note" section). These method is generally more laborious, particularly, with large samples, but available through S-plus, Analyse-it, EP Evaluator, and MedCalc and other software programs.

# 7 Note

More background, theoretical and mathematical information of validity assessments of diagnostic tests with continuous outcomes is given in Statistics applied to clinical studies 5th edition, the Chaps. 45 and 50, Springer Heidelberg Germany, 2012, from the same authors.

# Chapter 27
# Robust Tests for Imperfect Data

## 1 General Purpose

Robust tests are wonderful for imperfect data, because they often produce significant results, when standard tests don't. They may be able to handle major outliers in data files without largely changing the overall test results.

## 2 Schematic Overview of Type of Data File

| Outcome |
| --- |
| . |
| . |
| . |
| . |
| . |
| . |
| . |
| . |
| . |
| |
| |
| |
| |



T.J. Cleophas, A.H. Zwinderman, *Clinical Data Analysis on a Pocket Calculator*,
DOI 10.1007/978-3-319-27104-0_27

## 3 Primary Scientific Question

With data imperfect due to major outliers, can robust tests provide significant effects, if traditional tests don't.

## 4 Data Example

Frailty score-improvements after physiotherapy of 33 patients are measured in a study. The data are in the second column of underneath data.

| Patient | Score-improvement | Deviation from median | Trimmed data | Winsorized data |
|---|---|---|---|---|
| 1 | −8.00 | 11 | | −1.00 |
| 2 | −8.00 | 11 | | −1.00 |
| 3 | −8.00 | 7 | | −1.00 |
| 4 | −4.00 | 7 | | −1.00 |
| 5 | −4.00 | 7 | | −1.00 |
| 6 | −4.00 | 7 | | −1.00 |
| 7 | −4.00 | 7 | | −1.00 |
| 8 | −1.00 | 4 | −1.00 | −1.00 |
| 9 | 0.00 | 3 | 0.00 | 0.00 |
| 10 | 0.00 | 3 | 0.00 | 0.00 |
| 11 | 0.00 | 3 | 0.00 | 0.00 |
| 12 | 1.00 | 2 | 1.00 | 1.00 |
| 13 | 1.00 | 2 | 1.00 | 1.00 |
| 14 | 2.00 | 1 | 2.00 | 2.00 |
| 15 | 2.00 | 1 | 2.00 | 2.00 |
| 16 | 2.00 | 1 | 2.00 | 2.00 |
| 17 | 3.00 | median | 3.00 | 3.00 |
| 18 | 3.00 | 0 | 3.00 | 3.00 |
| 19 | 3.00 | 0 | 3.00 | 3.00 |
| 20 | 3.00 | 0 | 3.00 | 3.00 |
| 21 | 4.00 | 1 | 4.00 | 4.00 |
| 22 | 4.00 | 1 | 4.00 | 4.00 |
| 23 | 4.00 | 1 | 4.00 | 4.00 |
| 24 | 4.00 | 1 | 4.00 | 4.00 |
| 25 | 5.00 | 2 | 5.00 | 5.00 |
| 26 | 5.00 | 2 | 5.00 | 5.00 |
| 27 | 5.00 | 2 | | 5.00 |
| 28 | 5.00 | 2 | | 5.00 |
| 29 | 6.00 | 3 | | 5.00 |
| 30 | 6.00 | 3 | | 5.00 |
| 31 | 6.00 | 3 | | 5.00 |
| 32 | 7.00 | 4 | | 5.00 |
| 33 | 8.00 | 5 | | 5.00 |

The data suggest the presence of some central tendency: the values 3.00 and 5.00 are observed more frequently than the rest. However, the one sample t-test shows a mean difference from zero of 1.45 scores with a p-value of 0.067. Thus, not statistically significant.

# 5 T-Test for Medians and Median Absolute Deviations (MADs)

Underneath are descriptives of the above data that are appropriate for robust testing are given.

| | |
|---|---|
| Mean | 1.455 |
| standard deviation | 4.409 |
| standard error | 0.768 |
| mean after replacing outcome 1st patient with 0.00 | 1.697 |
| mean after replacing outcome first 3 patients with 0.00 | 2.182 |
| median | 3.000 |
| MAD | 2.500 |
| mean of the Winsorized data | 1.364 |
| standard deviation of the Winsorized data | 3.880 |

MAD = median absolute deviation = the median value of the sorted deviations from the median of a data file.

If the mean does not accurately reflect the central tendency of the data e.g. in case of outliers (highly unusual values), then the median (value in the middle) or the mode (value most frequently observed) may be a better alternative to summarizing the data and making predictions from them.

$$\text{Median} = 3.00$$

The above example shows in the third column the deviations from the median, and the table gives the median of the deviations from median (MAD = median absolute deviation).

$$\text{MAD} = 2.50$$

If we assume, that the data, though imperfect, are from a normal distribution, then the standard deviation of this normal distribution can be approximated from the equation

$$\text{standard deviation}_{\text{median}} = 1.426 \times \text{MAD} = 3.565$$

$$\text{standard error}_{\text{median}} = 3.565/\sqrt{n} = 3.565/\sqrt{33} = 0.6206$$

A t-test is, subsequently, performed, and produces a very significant effect: physiotherapy is really helpful.

$$t = \text{median}/\text{standard error}_{\text{median}} = 3.00/0.6206 = 4.834.$$

The underneath t-table is used for determining the p-value. For 32 (=33 − 1) degrees of freedom, and a t-value >3.646 means a p-value of <0.001 (two-tail).

| df | One-Tail = .4<br>Two-Tail = .8 | .25<br>.5 | .1<br>.2 | .05<br>.1 | .025<br>.05 | .01<br>.02 | .005<br>.01 | .0025<br>.005 | .001<br>.002 | .0005<br>.001 |
|---|---|---|---|---|---|---|---|---|---|---|
| 1 | 0.325 | 1.000 | 3.078 | 6.314 | 12.706 | 31.821 | 63.657 | 127.32 | 318.31 | 636.62 |
| 2 | 0.289 | 0.816 | 1.886 | 2.920 | 4.303 | 6.965 | 9.925 | 14.089 | 22.327 | 31.598 |
| 3 | 0.277 | 0.765 | 1.638 | 2.353 | 3.182 | 4.541 | 5.841 | 7.453 | 10.214 | 12.924 |
| 4 | 0.271 | 0.741 | 1.533 | 2.132 | 2.776 | 3.747 | 4.604 | 5.598 | 7.173 | 8.610 |
| 5 | 0.267 | 0.727 | 1.476 | 2.015 | 2.571 | 3.365 | 4.032 | 4.773 | 5.893 | 6.869 |
| 6 | 0.265 | 0.718 | 1.440 | 1.943 | 2.447 | 3.143 | 3.707 | 4.317 | 5.208 | 5.959 |
| 7 | 0.263 | 0.711 | 1.415 | 1.895 | 2.365 | 2.998 | 3.499 | 4.029 | 4.785 | 5.408 |
| 8 | 0.262 | 0.706 | 1.397 | 1.860 | 2.306 | 2.896 | 3.355 | 3.833 | 4.501 | 5.041 |
| 9 | 0.261 | 0.703 | 1.383 | 1.833 | 2.262 | 2.821 | 3.250 | 3.690 | 4.297 | 4.781 |
| 10 | 0.260 | 0.700 | 1.372 | 1.812 | 2.228 | 2.764 | 3.169 | 3.581 | 4.144 | 4.587 |
| 11 | 0.260 | 0.697 | 1.363 | 1.796 | 2.201 | 2.718 | 3.106 | 3.497 | 4.025 | 4.437 |
| 12 | 0.259 | 0.695 | 1.356 | 1.782 | 2.179 | 2.681 | 3.055 | 3.428 | 3.930 | 4.318 |
| 13 | 0.259 | 0.694 | 1.350 | 1.771 | 2.160 | 2.650 | 3.012 | 3.372 | 3.852 | 4.221 |
| 14 | 0.258 | 0.692 | 1.345 | 1.761 | 2.145 | 2.624 | 2.977 | 3.326 | 3.787 | 4.140 |
| 15 | 0.258 | 0.691 | 1.341 | 1.753 | 2.131 | 2.602 | 2.947 | 3.286 | 3.733 | 4.073 |
| 16 | 0.258 | 0.690 | 1.337 | 1.746 | 2.120 | 2.583 | 2.921 | 3.252 | 3.686 | 4.015 |
| 17 | 0.257 | 0.689 | 1.333 | 1.740 | 2.110 | 2.567 | 2.898 | 3.222 | 3.646 | 3.965 |
| 18 | 0.257 | 0.688 | 1.330 | 1.734 | 2.101 | 2.552 | 2.878 | 3.197 | 3.610 | 3.922 |
| 19 | 0.257 | 0.688 | 1.328 | 1.729 | 2.093 | 2.539 | 2.861 | 3.174 | 3.579 | 3.883 |
| 20 | 0.257 | 0.687 | 1.325 | 1.725 | 2.086 | 2.528 | 2.845 | 3.153 | 3.552 | 3.850 |
| 21 | 0.257 | 0.686 | 1.323 | 1.721 | 2.080 | 2.518 | 2.831 | 3.135 | 3.527 | 3.819 |
| 22 | 0.256 | 0.686 | 1.321 | 1.717 | 2.074 | 2.508 | 2.819 | 3.119 | 3.505 | 3.792 |
| 23 | 0.256 | 0.685 | 1.319 | 1.714 | 2.069 | 2.500 | 2.807 | 3.104 | 3.485 | 3.767 |
| 24 | 0.256 | 0.685 | 1.318 | 1.711 | 2.064 | 2.492 | 2.797 | 3.091 | 3.467 | 3.745 |
| 25 | 0.256 | 0.684 | 1.316 | 1.708 | 2.060 | 2.485 | 2.787 | 3.078 | 3.450 | 3.725 |
| 26 | 0.256 | 0.684 | 1.315 | 1.706 | 2.056 | 2.479 | 2.779 | 3.067 | 3.435 | 3.707 |
| 27 | 0.256 | 0.684 | 1.314 | 1.703 | 2.052 | 2.473 | 2.771 | 3.057 | 3.421 | 3.690 |
| 28 | 0.256 | 0.683 | 1.313 | 1.701 | 2.048 | 2.467 | 2.763 | 3.047 | 3.408 | 3.674 |
| 29 | 0.256 | 0.683 | 1.311 | 1.699 | 2.045 | 2.462 | 2.756 | 3.038 | 3.396 | 3.659 |
| 30 | 0.256 | 0.683 | 1.310 | 1.697 | 2.042 | 2.457 | 2.750 | 3.030 | 3.385 | 3.646 |
| 40 | 0.255 | 0.681 | 1.303 | 1.684 | 2.021 | 2.423 | 2.704 | 2.971 | 3.307 | 3.551 |
| 60 | 0.254 | 0.679 | 1.296 | 1.671 | 2.000 | 2.390 | 2.660 | 2.915 | 3.232 | 3.460 |
| 120 | 0.254 | 0.677 | 1.289 | 1.658 | 1.980 | 2.358 | 2.617 | 2.860 | 3.160 | 3.373 |
| ∞ | 0.253 | 0.674 | 1.282 | 1.645 | 1.960 | 2.326 | 2.576 | 2.807 | 3.090 | 3.291 |

The t-table has a left-end column giving degrees of freedom (≈ sample sizes), and two top rows with p-values (areas under the curve = p − values), one-tail meaning that only one end of the curve, two-tail meaning that both ends are assessed simultaneously. The t-table is, furthermore, full of t-values, that, with ∞ degrees of freedom, are equal to z-values (Chap. 36). The t-values are to be understood as mean results of studies, but not expressed in mmol/l, kilograms, but in so-called SEM-units (Standard error of the mean units), that are obtained by dividing your mean result by its own standard error. With many degrees of freedom (large samples) the curve will be a little bit narrower, and more in agreement with nature.

## 6 T-Test for Winsorized Variances

The terminology comes from Winsor's principle: all observed distributions are Gaussian in the middle. First, we have to trim the data, e.g., by 20 % on either side (see data file, fourth column). Then, we have to fill up their trimmed values with Winsorized scores, which are the smallest and largest untrimmed scores (data file, fifth column). The mean is, then, calculated, as well as the standard deviation and standard error, and a t-test is performed for null hypothesis testing.

Winsorized mean = 1.364
Winsorized standard deviation = 3.880
Winsorized standard error = 3.880 / √n
= 3.880 / √ 33
= 0.675

t-test
t = Winsorized mean / Winsorized standard error
t = 2.021

With 32 (=33−1) degrees of freedom and a t-value > 2.2042 means a p-value < 0.05 (two-sided).

p-value < 0.05

## 7 Mood's Test (One Sample Wilcoxon's Test)

| | −8.00 | −8.00 | −8.00 | −4.00 | −4.00 | −4.00 | −1.00 | 0.00 | 0.00 | 0.00 | 1.00 | …. |
|---|---|---|---|---|---|---|---|---|---|---|---|---|
| −8.00 | | −8.00 | −8.00 | −6.00 | −6.00 | −6.00 | −4.50 | −4.00 | −4.00 | −4.00 | −3.50 | |
| −8.00 | | | −8.00 | −6.00 | −6.00 | −6.00 | −4.50 | −4.00 | −4.00 | −4.00 | −3.50 | |
| −8.00 | | | | −6.00 | −6.00 | −6.00 | −4.50 | −4.00 | −4.00 | −4.00 | −3.50 | |
| −4.00 | | | | | −4.00 | −4.00 | −2.50 | −2.00 | −2.00 | −2.00 | −1.50 | |
| −4.00 | | | | | | −4.00 | −2.50 | −2.00 | −2.00 | −2.00 | −1.50 | |
| −4.00 | | | | | | | −2.50 | −2.00 | −2.00 | −2.00 | −1.50 | |
| −1.00 | | | | | | | | −0.50 | −0.50 | −0.50 | 0.00 | |
| 0.00 | | | | | | | | | 0.00 | 0.00 | 0.50 | |
| 0.00 | | | | | | | | | | 0.00 | 0.50 | |
| 0.00 | | | | | | | | | | | 0.50 | |
| 1.00 | | | | | | | | | | | | |
| …. | | | | | | | | | | | | |
| …. | | | | | | | | | | | | |

The Mood's test is, sometimes, called the one sample Wilcoxon's test. The above table shows how it works. Paired averages [(vertical value + horizontal value)/2] are calculated. If the data are equally distributed around an average of 0, then we will have half of the average being positive, half negative.

We observe 1122 paired averages,
1122/2 = 561 should be positive,
349 positive paired averages are found.

A chi-square test is performed
chi-square value = (Observed − expected numbers)$^2$/Expected numbers
chi-square value = $(349–561)^2 / 349 = 128.729$
$p < 0.001$ with 1 degree of freedom

The underneath chi-square table has an upper row with areas under the curve, a left-end column with degrees of freedom, and a whole lot of chi-square values. It shows that, for 1 degrees of freedom, and chi-square values >10.827, we will find a p-value <0.001.

Chi-squared distribution

| | Two-tailed *P*-value | | | |
|---|---|---|---|---|
| *df* | 0.10 | 0.05 | 0.01 | 0.001 |
| 1 | 2.706 | 3.841 | 6.635 | 10.827 |
| 2 | 4.605 | 5.991 | 9.210 | 13.815 |
| 3 | 6.251 | 7.851 | 11.345 | 16.266 |
| 4 | 7.779 | 9.488 | 13.277 | 18.466 |
| 5 | 9.236 | 11.070 | 15.086 | 20.515 |
| 6 | 10.645 | 12.592 | 16.812 | 22.457 |
| 7 | 12.017 | 14.067 | 18.475 | 24.321 |
| 8 | 13.362 | 15.507 | 20.090 | 26.124 |
| 9 | 14.684 | 16.919 | 21.666 | 27.877 |
| 10 | 15.987 | 18.307 | 23.209 | 29.588 |
| 11 | 17.275 | 19.675 | 24.725 | 31.264 |
| 12 | 18.549 | 21.026 | 26.217 | 32.909 |
| 13 | 19.812 | 22.362 | 27.688 | 34.527 |
| 14 | 21.064 | 23.685 | 29.141 | 36.124 |
| 15 | 22.307 | 24.996 | 30.578 | 37.698 |
| 16 | 23.542 | 26.296 | 32.000 | 39.252 |
| 17 | 24.769 | 27.587 | 33.409 | 40.791 |
| 18 | 25.989 | 28.869 | 34.805 | 42.312 |
| 19 | 27.204 | 30.144 | 36.191 | 43.819 |
| 20 | 28.412 | 31.410 | 37.566 | 45.314 |
| 21 | 29.615 | 32.671 | 38.932 | 46.796 |
| 22 | 30.813 | 33.924 | 40.289 | 48.268 |
| 23 | 32.007 | 35.172 | 41.638 | 49.728 |

(continued)

|  | Two-tailed *P*-value | | | |
|---|---|---|---|---|
| *df* | 0.10 | 0.05 | 0.01 | 0.001 |
| 24 | 33.196 | 36.415 | 42.980 | 51.179 |
| 25 | 34.382 | 37.652 | 44.314 | 52.619 |
| 26 | 35.536 | 38.885 | 45.642 | 54.051 |
| 27 | 36.741 | 40.113 | 46.963 | 55.475 |
| 28 | 37.916 | 41.337 | 48.278 | 56.892 |
| 29 | 39.087 | 42.557 | 49.588 | 58.301 |
| 30 | 40.256 | 43.773 | 50.892 | 59.702 |
| 40 | 51.805 | 55.758 | 63.691 | 73.403 |
| 50 | 63.167 | 67.505 | 76.154 | 86.660 |
| 60 | 74.397 | 79.082 | 88.379 | 99.608 |
| 70 | 85.527 | 90.531 | 100.43 | 112.32 |
| 80 | 96.578 | 101.88 | 112.33 | 124.84 |
| 90 | 107.57 | 113.15 | 124.12 | 137.21 |
| 100 | 118.50 | 124.34 | 135.81 | 149.45 |

## 8 Conclusion

The above three robust tests produced p-values of <0.001, <0.05, and <0.001, while the one sample t-test was not statistically significant. Robust tests are wonderful for imperfect data, because they often produce significant results, when standard tests don't.

## 9 Note

More background, theoretical and mathematical information of robust tests is given in SPSS for Starters part 2, Chap. 20, Springer Heidelberg Germany, 2012, from the same authors.

# Chapter 28
# Non-linear Modeling on a Pocket Calculator

## 1 General Purpose

Non-linear relationships in clinical research are often linear after logarithmic transformations. Odds ratios, log likelihood ratios, Markov models and many regression models are models that make use of it. An example with real data is given. We have to add that logarithmic transformation is not always successful, and that alternative methods are available like Box Cox transformation, and computationally intensive methods like spline and Loess modeling (see Chap. 24. In: Statistics Applied to Clinical Studies, Springer New York, 5th edition, 2012, and Chap. 14 of SPSS for Starters Part 2, Springer New York, 2012, both from the same authors). However, these methods generally require statistical software and can not be executed on a pocket calculator. This chapter assesses simply logarithmic transformation of the outcome variable for linearization of survival data.

## 2 Schematic Overview of Type of Data File

| Outcome | Predictor |
|---|---|
| . | . |
| . | . |
| . | . |
| . | . |
| . | . |
| . | . |
| . | . |
| . | . |

(continued)



T.J. Cleophas, A.H. Zwinderman, *Clinical Data Analysis on a Pocket Calculator*,
DOI 10.1007/978-3-319-27104-0_28

| Outcome | Predictor |
|---|---|
| . | . |
| | |
| | |
| | |
| | |
| | |

## 3 Primary Scientific Question

Can logarithmic transformation of survival data linearize survival patterns.

## 4 Data Example

The underneath figure shows the survivals of 240 patients with small cell carcinomas.

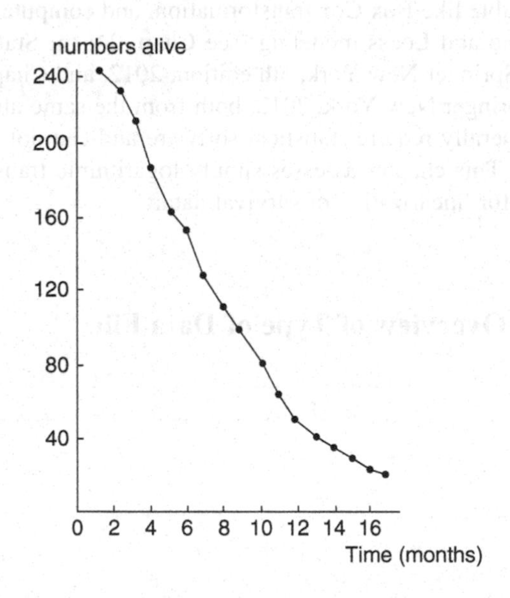


The underneath figure shows the natural logarithms of these survivals. It can be observed that logarithmic transformation of the numbers of patients alive readily produces a close to linear pattern.

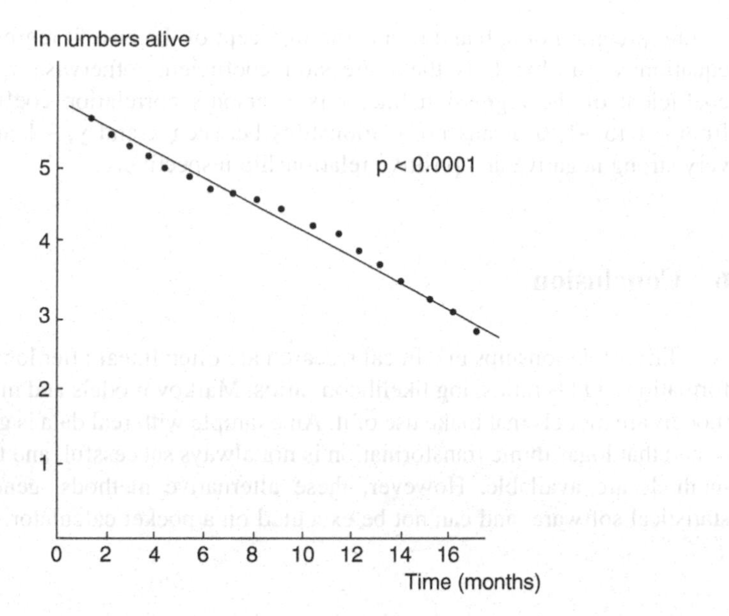


The equation of the above regression line of the data y = a + bx with a = intercept and b = regression coefficient can be calculated from a pocket calculator.

# 5 Calculation of Linear Regression Parameters from a Pocket Calculator

Some pocket calculators offer linear regression. An example is given (see also Chap. 8).

| x-values | y-values |
|---|---|
| Temp (°C) | Atmospheric pressure (hpa) |
| 10 | 1003 |
| 15 | 1005 |
| 20 | 1010 |
| 25 | 1011 |
| 30 | 1014 |

Electronic Calculator (see Chap. 1) can be used for the purpose.

Press:

on. . ..mode. . ..3. . ..1. . ..10. . ., . . ..1003. . ..M+ . . ..15. . .., . . .. 1005. . .. M+ . . ...
etc. . ....M+....shift....s-var....▶....▶....1.....a is given. . ..shift....s-var ....▶....▶
....2.....b is given. . .. shift....s-var....▶....▶....3.....r is given. . ..

Interpretation of a, b and r.: a is the intercept of the best fit regression line with equation $y = a + bx$; b is the regression coefficient, otherwise called direction coefficient of the regression line; r is Pearson's correlation coefficient, it runs from −1 to +1, 0 means no relationships between x and y, −1 and +1 mean a very strong negative and positive relationship respectively.

## 6 Conclusion

Non-linear relationships in clinical research are often linear after logarithmic transformations. Odds ratios, log likelihood ratios, Markov models and many regression models are models that make use of it. An example with real data is given. We have to add that logarithmic transformation is not always successful, and that alternative methods are available. However, these alternative methods, generally, require statistical software, and can not be executed on a pocket calculator.

## 7 Note

More background, theoretical and mathematical information is given in Statistics Applied to Clinical Studies, Springer New York, 5th edition, Chap. 24, 2012, and SPSS for Starters Part 2, Springer New York, 2012, Chap. 14, both from the same authors.

# Chapter 29
# Fuzzy Modeling for Imprecise and Incomplete Data

## 1 General Purpose

Fuzzy modeling is a methodology that works with partial truths: it can answer questions to which the answers are "yes" and "no" at different times or partly "yes" and "no" at the same time. It can be used to match any type of data, particularly incomplete and imprecise data, and it is able to improve precision of such data. It can be applied with any type of statistical distribution and it is, particularly, suitable for uncommon and unexpected non linear relationships. This chapter assesses the use of fuzzy modeling of clinical data.

## 2 Schematic Overview of Type of Data File

| Outcome |
|---|
| . |
| . |
| . |
| . |
| . |
| . |
| . |
| . |
| . |





T.J. Cleophas, A.H. Zwinderman, *Clinical Data Analysis on a Pocket Calculator*,
DOI 10.1007/978-3-319-27104-0_29

## 3 Primary Scientific Question

Is fuzzy modeling able to provide improved precision of imprecise clinical data.

## 4 Fuzzy Terms

### 4.1 Universal Space

Defined range of input values, defined range of output values.

### 4.2 Fuzzy Memberships

The universal spaces are divided into equally sized parts called membership functions

### 4.3 Linguistic Membership Names

Each fuzzy membership is given a name, otherwise called linguistic term.

### 4.4 Triangular Fuzzy Sets

A common way of drawing the membership function with on the x-axis the input values, on the y-axis the membership grade for each input value.

### 4.5 Fuzzy Plots

Graphs summarizing the fuzzy memberships of (for example) the input values.

### 4.6 Linguistic Rules

The relationships between the fuzzy memberships of the input data and those of the output data (the method of calculation is shown in the underneath examples).

# 5 Data Example 1

Underneath are the quantal pharmacodynamic effects of different induction dosages of thiopental on numbers of responding subjects.

| Input values induction dosage of thiopental (mg/kg) | output values numbers of responders (n) | fuzzy-modeled output numbers of responders (n) |
|---|---|---|
| 1 | 4 | 4 |
| 1. | 5 | 5 |
| 2 | 6 | 8 |
| 2.5 | 9 | 10 |
| 3 | 12 | 12 |
| 3.5 | 17 | 14 |
| 4 | 17 | 16 |
| 4.5 | 12 | 14 |
| 5 | 9 | 12 |

The effects of different induction dosages of thiopental on numbers of responding subjects are in the above table, left two columns. The right column gives the fuzzy-modeled output. The figures below show that the un-modeled curve (upper curve) fits the data less well than does the modeled (lower curve).

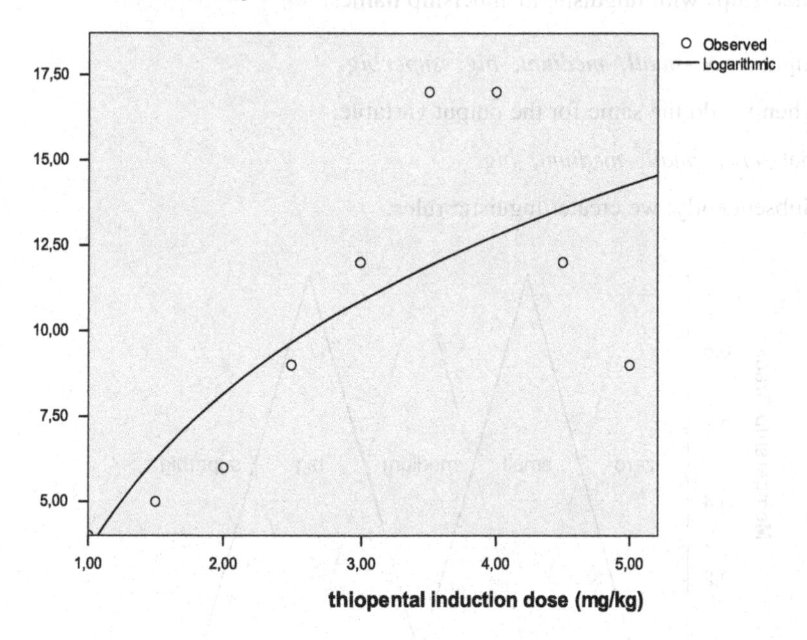

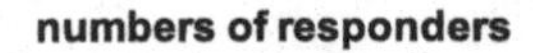


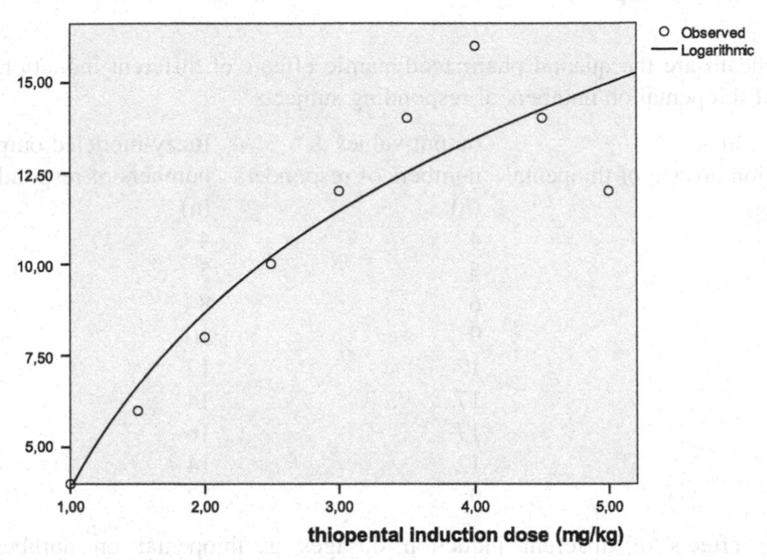


We fuzzy-model the input and output relationships (figures below).

First of all, we create linguistic rules for the input and output data.

For that purpose we divide the universal space of the input variable into fuzzy memberships with linguistic membership names:

input-*zero, -small, -medium, -big, -superbig.*

Then we do the same for the output variable:

output-*zero, -small, -medium, -big.*

Subsequently, we create linguistic rules.

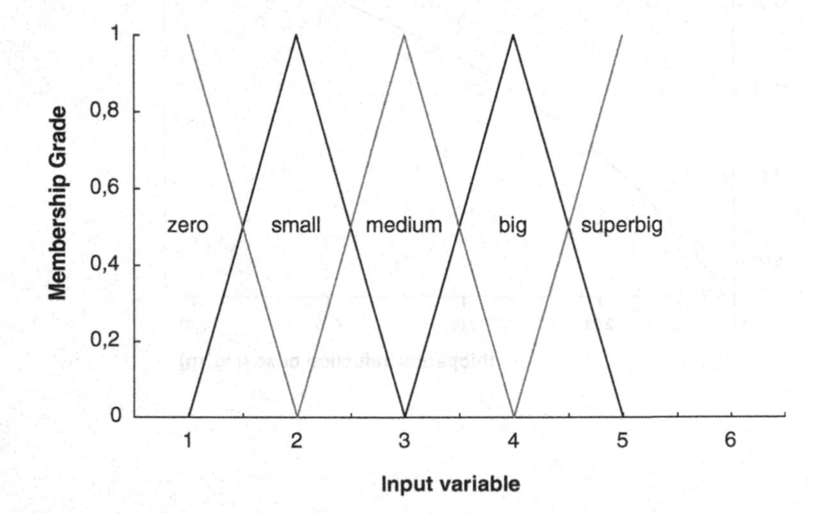

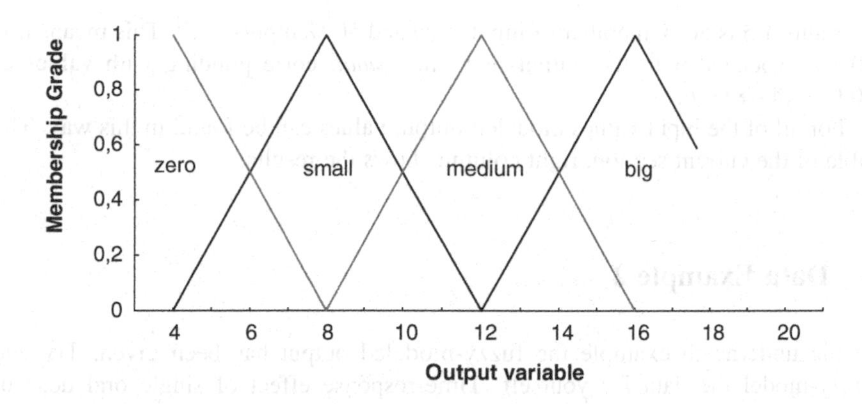


The above figure shows that input-*zero* consists of the values 1 and 1.5.

The value 1 (100 % membership) has 4 as outcome value (100 % membership of output-*zero*).
The value 1.5 (50 % membership) has 5 as outcome value (75 % membership of output-*zero*, 25 % of output-*small*).

The input-*zero* produces 100%x100% + 50%x75% = 137.5 % membership to output-*zero*, and 50% × 25% = 12.5 % membership to output-*small*, and so, output-*zero* is the most important output contributor here, and we forget about the small contribution of output-*small*.

Input-*small* is more complex, it consists of the values 1.5, and 2.0, and 2.5.

The value 1.5 (50 % membership) has 5 as outcome value (75 % membership of output-*zero*, 25 % membership of output-*small*).
The value 2.0 (100 % membership) has 6 as outcome value (50 % membership of outcome-*zero*, and 50 % membership of output-*small*).
The value 2.5 (50 % membership) has 9 as outcome value (75 % membership of output-*small* and 25 % of output-*medium*).

The input-*small* produces 50 % × 75 % + 100 % × 50 % = 87.5 % membership to output-*zero*, 50 % × 25 % + 100 % × 50 % + 50 % × 75 % = 100 % membership to output-*small*, and 50 % × 25 % = 12.5 % membership to output-*medium*. And so, the output-*small* is the most important contributor here, and we forget about the other two.

For the other input memberships similar linguistic rules are determined:

Input-*medium* → output-*medium*
Input-*big* → output-*big*
Input-*superbig* → output-*medium*

We are, particularly interested in the modeling capacity of fuzzy logic in order to improve the precision of pharmacodynamic modeling.

The modeled output value of input value 1 is found as follows.

Value 1 is 100 % member of input-*zero*, meaning that according to the above linguistic rules it is also associated with a 100 % membership of output-*zero* corresponding with a value of 4.

Value 1.5 is 50 % member of input-*zero* and 50 % input-*small*. This means it is 50 % associated with the output-*zero* and *–small* corresponding with values of 50 % × (4 + 8) = 6.

For all of the input values modeled output values can be found in this way. The table of the current section, right column shows the results.

# 6 Data Example 2

In the underneath example the fuzzy-modeled output has been given. Try and fuzzy-model the data for yourself. Time-response effect of single oral dose of 120 mg propranolol on peripheral arterial flow.

| Input values<br>Hours after oral Administration of 120 mg propranolol | output values<br>peripheral arterial flow (ml/100 ml tissue/min) | fuzzy-modeled output<br>peripheral arterial flow (ml/100 ml tissue/min) |
|---|---|---|
| 1 | 20 | 20 |
| 2 | 12 | 14 |
| 3 | 9 | 8 |
| 4 | 6 | 6 |
| 5 | 5 | 4 |
| 6 | 4 | 4 |
| 7 | 5 | 4 |
| 8 | 6 | 6 |
| 9 | 9 | 8 |
| 10 | 12 | 14 |
| 11 | 20 | 20 |

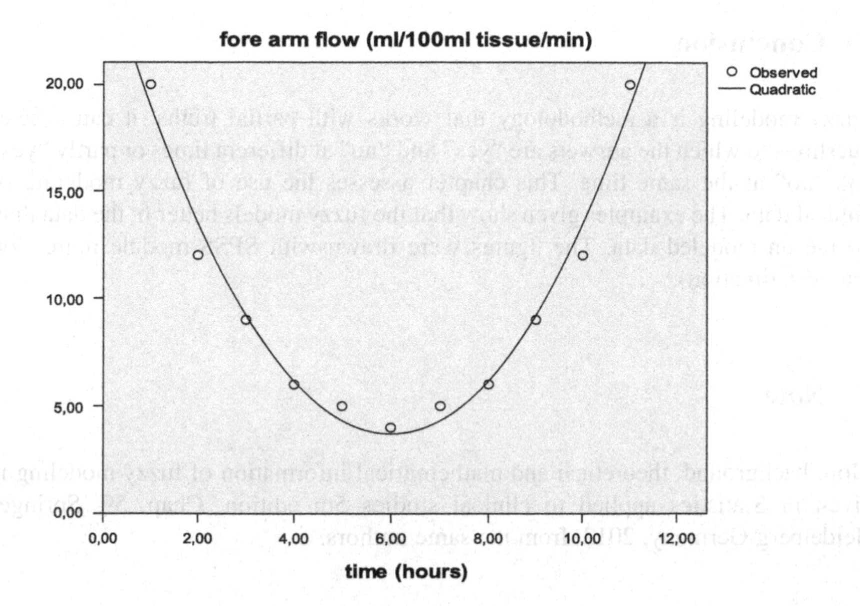


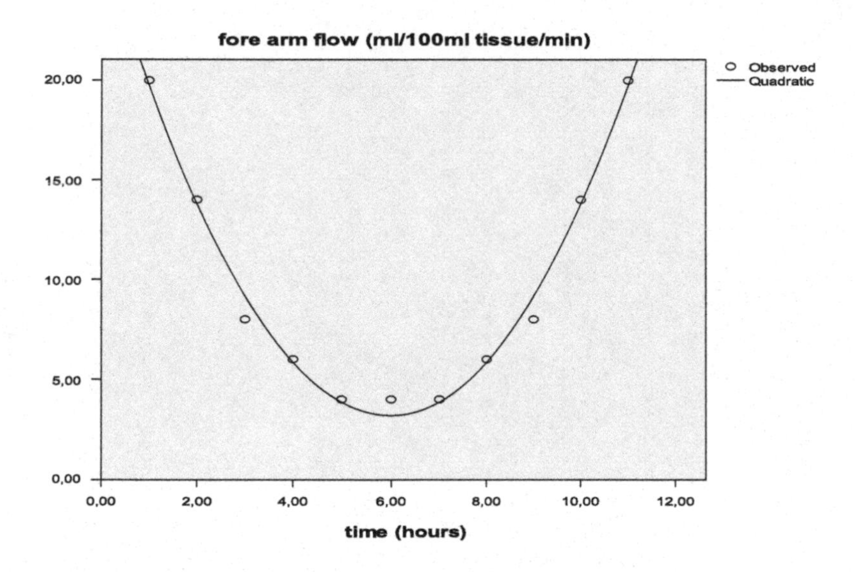


Pharmacodynamic relationship between the time after oral administration of 120 mg of propranolol (x-axis, hours) and absolute change in fore arm flow (y-axis, ml/100 ml tissue/min) are in the above graphs. The un-modeled curve (upper curve) fits the data slightly less well than does the modeled (lower curve).

## 7 Conclusion

Fuzzy modeling is a methodology that works with partial truths: it can answer questions to which the answers are "yes" and "no" at different times or partly "yes" and "no" at the same time. This chapter assesses the use of fuzzy modeling of clinical data. The examples given show that the fuzzy models better fit the data than do the un-modeled data. The figures were drawn with SPSS module regression (curve estimation).

## 8 Note

More background, theoretical and mathematical information of fuzzy modeling is given in Statistics applied to clinical studies 5th edition, Chap. 59, Springer Heidelberg Germany, 2012, from the same authors.

# Chapter 30
# Bhattacharya Modeling for Unmasking Hidden Gaussian Curves

## 1 General Purpose

Bhattacharya modeling can be used for unmasking Gaussian curves in the data. It should with the help of log transformed frequency scores of data histograms enable to identify Gaussian subsets in the data. It can also be applied to produce a better Gaussian fit to a data file than the usual mean and standard deviation does. This chapter assesses how it can be used to identify Gaussian data subsets, and provide models better fitting the data, than the traditional methods do.

## 2 Schematic Overview of Type of Data File

| Outcome |
|---|
| . |
| . |
| . |
| . |
| . |
| . |
| . |
| . |
| . |
| |
| |
| |



T.J. Cleophas, A.H. Zwinderman, *Clinical Data Analysis on a Pocket Calculator*,
DOI 10.1007/978-3-319-27104-0_30

## 3 Primary Scientific Question

Can Bhattacharya modeling be applied for determining normal values of diagnostic tests and their confidence intervals, and for searching subsets in the data.

## 4 Data Example

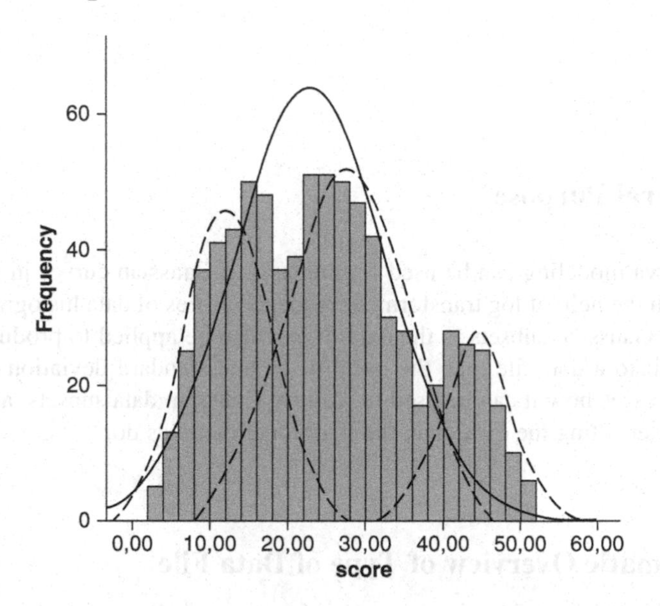


The above graph gives an example of the frequency distributions of vascular lab scores of a population of 787 patients at risk of peripheral vascular disease. The continuous Gaussian curves are calculated from the mean ± standard deviation, the interrupted Gaussian curves from Bhattacharya modeling. The pattern of the histogram is suggestive of certain subsets in this population. The underneath table left two columns give the scores and frequencies. The frequencies are log (logarithmic) transformed (third column) (see also 27 and 43), and, then, the differences between two subsequent log transformed scores are calculated (fourth column).

| Score | frequency | log | delta log |
|---|---|---|---|
| 2 | 1 | 0.000 | 0.000 |
| 4 | 5 | 0.699 | 0.699 |
| 6 | 13 | 1.114 | 0.415 |
| 8 | 25 | 1.398 | 0.284 |
| 10 | 37 | 1.568 | 0.170 |
| 12 | 41 | 1.613 | 0.045 |
| 14 | 43 | 1.633 | 0.020 |

| | | | |
|---|---|---|---|
| 16 | 50 | 1.699 | −0.018 |
| 18 | 48 | 1.681 | −0.111 |
| 20 | 37 | 1.570 | 0.021 |
| 22 | 39 | 1.591 | 0.117 |
| 24 | 51 | 1.708 | 0.000 |
| 26 | 51 | 1.708 | −0.009 |
| 28 | 50 | 1.699 | −0.027 |
| 30 | 47 | 1.672 | −0.049 |
| 32 | 42 | 1.623 | −0.146 |
| 34 | 30 | 1.477 | −0.176 |
| 36 | 28 | 1.447 | −0.030 |
| 38 | 16 | 1.204 | −0.243 |
| 40 | 20 | 1.301 | 0.097 |
| 42 | 28 | 1.447 | 0.146 |
| 44 | 26 | 1.415 | −0.032 |
| 46 | 25 | 1.398 | −0.017 |
| 48 | 17 | 1.230 | −0.168 |
| 50 | 10 | 1.000 | −0.230 |
| 52 | 6 | 0.778 | −0.222 |

The underneath graph shows the plot of the scores against the delta log terms. Three straight lines are identified. The equations of these lines can be calculated (using linear regression) or extrapolated from the graph below.

1. $y = 0.944 - 0.078\ x$
2. $y = 0.692 - 0.026\ x$
3. $y = 2.166 - 0.048\ x$

The characteristics of the corresponding Gaussian curves can be calculated as follows.

1. mean = −0.944/−0.078 = 12.10, standard deviation = 1/0.078 = 12.82
2. mean = −0.692/−0.026 = 26.62, standard deviation = 1/0.026 = 38.46
3. mean = −2.166/−0.048 = 45.13, standard deviation = 1/0.048 = 20.83.

In the underneath graph the above three Gaussian curves as given are drawn as interrupted curves. Here the scores from the above graph have been plotted against the delta log terms as calculated from the frequencies from the above graph.

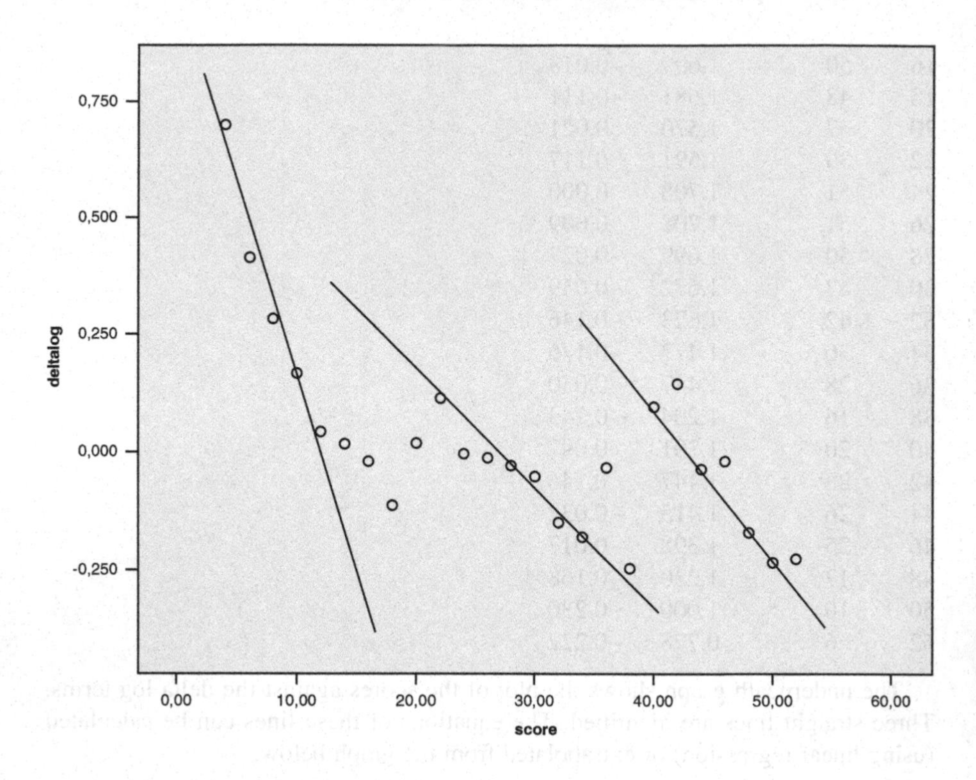


## 5 Conclusion

We conclude that Bhattacharya modeling with the help of log transformed frequency scores of data histograms enables to identify Gaussian subsets in the data. It can also be applied to produce a better Gaussian fit to a data file than the usual mean and standard deviation does (Chapter 59. In: Statistics Applied to Clinical Studies, Springer New York, 5th edition, 2012, from the same authors).

## 6 Note

More background, theoretical and mathematical information of Bhattacharya modeling is given in the Statistics applied to clinical studies 5th edition, Chap. 26 and 59, Springer Heidelberg Germany, 2012, from the same authors.

# Chapter 31
# Item Response Modeling Instead of Classical Linear Analysis of Questionnaires

## 1 General Purpose

Item response modeling is used for analyzing psychometric data, quality of life data, and can even be used analyzing diagnostic tests. It may provide better precision than does the classical linear analysis. This chapter assesses how it works.

## 2 Schematic Overview of Type of Data File

| Outcome (answer pattern) |
|---|
| . |
| . |
| . |
| . |
| . |
| . |
| . |
| . |
| . |
| |
| |
| |
| |



T.J. Cleophas, A.H. Zwinderman, *Clinical Data Analysis on a Pocket Calculator*,
DOI 10.1007/978-3-319-27104-0_31

## 3 Primary Scientific Question

Are item response models more sensitive than classical linear methods for making predictions from psychological/QOL questionnaires, and diagnostic tests.

## 4 Example

One of the estimators of quality of life (QOL) is "feeling happy". Five yes/no questions indicate increasing levels of daily happiness: (1) 0 hours happy, (2) 6 hours happy, (3) 12 hours happy, (4) 18 hours happy, (5) 24 hours happy. Usually, with five yes/no questions in a domain the individual result is given in the form of a score, here, e.g., a score from 0 to 5 dependent on the number of positive answers given per person. However, with many questionnaires different questions represent different levels of difficulty or different levels of benefit etc. This can be included in the analysis using item response modeling.

A summary of a 5-item quality of life data of 1000 anginal patients is given below.

| No. response pattern | Response pattern (1 = yes, 2 = no) to items 1 to 5 | Observed Frequencies |
|---|---|---|
| 1. | 11111 | 0 |
| 2. | 11112 | 0 |
| 3. | 11121 | 0 |
| 4. | 11122 | 1 |
| 5. | 11211 | 2 |
| 6. | 11212 | 3 |
| 7. | 11221 | 5 |
| 8. | 11222 | 8 |
| 9. | 12111 | 12 |
| 10. | 12112 | 15 |
| 11. | 12121 | 18 |
| 12. | 12122 | 19 |
| 13. | 12211 | 20 |
| 14. | 12212 | 21 |
| 15. | 12221 | 21 |
| 16. | 12222 | 21 |
| 17. | 21111 | 20 |
| 18. | 21112 | 19 |
| 19. | 21121 | 18 |
| 20. | 21122 | 15 |
| 21. | 21211 | 12 |
| 22. | 21212 | 9 |

| | | |
|---|---|---|
| 23. | 21221 | 5 |
| 24. | 21222 | 3 |
| 25. | 22111 | 4 |
| 26. | 22112 | 0 |
| 27. | 22121 | 0 |
| 28. | 22122 | 0 |
| 29. | 22211 | 0 |
| 30. | 22212 | 0 |
| 31. | 22221 | 0 |
| 32. | 22222 | 0_____+ |
| | | 271 |

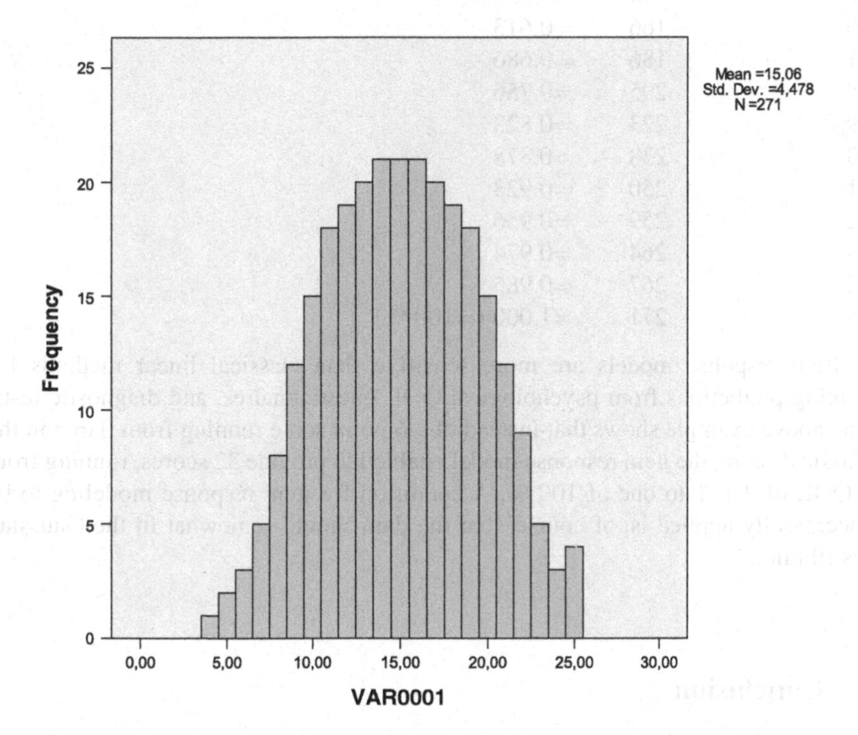


The above table shows how 5 questions can be used to produce 32 different answer patterns. The above graph figure shows a histogram of the answer patterns with the type of pattern on the x-axis and "how often" n the y-axis. A Gaussian like distribution frequency is observed. A score around 15 is observed most frequently, and can be interpreted as the mean score of the study. Low scores indicate little QOL. High scores indicate high QOL. Underneath the areas under the curve (AUCs) of the histogram is also given. The larger the AUCs, which run from 0.004 to 1.000 (0.4–100 %), the better the QOL.

| Response pattern | AUC (area under the curve) | |
|---|---|---|
| 4 | 1/271 | =0.004 (=4 %) |
| 5 | 3 | =0.011 |
| 6 | 6 | =0.022 |
| 7 | 11 | =0.041 |
| 8 | 19 | =0.070 |
| 9 | 31 | =0.114 |
| 10 | 46 | =0.170 |
| 11 | 64 | =0.236 |
| 12 | 83 | =0.306 |
| 13 | 103 | =0.380 |
| 14 | 124 | =0.458 |
| 15 | 145 | =0.535 |
| 16 | 166 | =0.613 |
| 17 | 186 | =0.686 |
| 18 | 205 | =0.756 |
| 19 | 223 | =0.823 |
| 20 | 238 | =0.878 |
| 21 | 250 | =0.923 |
| 22 | 259 | =0.956 |
| 23 | 264 | =0.974 |
| 24 | 267 | =0.985 |
| 25 | 271 | =1.000 (=100 %) |

Item response models are more sensitive than classical linear methods for making predictions from psychological/QOL questionnaires, and diagnostic tests. The above example shows that instead of a 6 point score running from 0 to 5 in the classical score, the item response model enabled to provide 32 scores, running from a QOL of 0.4 % to one of 100 %. A condition for item response modeling to be successfully applied is, of course, that the data should somewhat fit the Gaussian distribution.

## 5 Conclusion

Item response models are more sensitive than classical linear methods for making predictions from psychological/QOL questionnaires, and diagnostic tests.

Item response modeling is not in SPSS, but the LTA-2 software program of Uebersax is a free software program for the purpose. It works with the areas under the curve of statistically modeled best fit Gaussian curves of the data rather than a histogram of the data, but, otherwise, it is similar to the pocket calculator method.

# 6 Note

More background, theoretical and mathematical information of item response modeling is given in Statistics applied to clinical studies 5th edition, Chap. 39, Springer Heidelberg Germany, 2012, from the same authors.

# Chapter 32
# Meta-analysis of Continuous Data

## 1 General Purpose

Meta-analyses can be defined as systematic reviews with pooled data. Because the separate studies in a meta-analysis have different sample sizes for the overall results a weighted average has to be calculated. Heterogeneity in a meta-analysis means that the differences in the results between the studies are larger than could happen by chance. The calculation of the overall result and the test for heterogeneity is demonstrated underneath.

## 2 Schematic Overview of Type of Data File

| Outcome | Predictor | Patient characteristic |
|---|---|---|
| . | . | . |
| . | . | . |
| . | . | . |
| . | . | . |
| . | . | . |
| . | . | . |
| . | . | . |
| . | . | . |
| | | |
| | | |
| | | |
| | | |
| | | |
| | | |
| | | |

 
T.J. Cleophas, A.H. Zwinderman, *Clinical Data Analysis on a Pocket Calculator*,
DOI 10.1007/978-3-319-27104-0_32

## 3 Primary Scientific Question

How do we assess pooled results of multiple studies, how do we assess heterogeneity between the studies.

## 4 Data Example, Pooling

A meta-analysis of the difference in systolic blood pressures (mm Hg) between patients treated with potassium and those with placebo. Diff = difference in systolic blood pressure between patients on potassium and placebo, var = variance = (standard error)$^2$

| | N | diff (systolic) | standard error | 1/var | diff/var | diff$^2$/var |
|---|---|---|---|---|---|---|
| 1. McGregor 1982 | 23 | −7.0 | 3.1 | 0.104 | −0.728 | 5.096 |
| 2. Siani 1987 | 37 | −14.0 | 4.0 | 0.063 | −0.875 | 12.348 |
| 3. Svetkey 1987 | 101 | −6.4 | 1.9 | 0.272 | −1.773 | 11.346 |
| 4. Krishna 1989 | 10 | −5.5 | 3.8 | 0.069 | −0.380 | 2.087 |
| 5. Obel 1989 | 48 | −41.0 | 2.6 | 0.148 | −6.065 | 248.788 |
| 6. Patki 1990 | 37 | −12.1 | 2.6 | 0.148 | −1.791 | 21.669 |
| 7. Fotherby 1992 | 18 | −10.0 | 3.8 | 0.069 | −0.693 | 6.900 |
| 8. Brancati 1996 | 87 | −6.9 | 1.2 | 0.694 | −4.792 | 33.041 |
| 9. Gu 2001 | 150 | −5.0 | 1.4 | 0.510 | −2.551 | 12.750 |
| 10. Sarkkinen 2011 | 45 | −11.3 | 4.8 | 0.043 | −0.490 | 5.091 |
| | | | | | | + |
| | | | | 2.125 | −20.138 | 359.516 |

Pooled difference $= -20.138/2.125 = -9.48$ mm Hg
Chi-square value for pooled data $= (-20.138)^2 / 2.125 = 206.91$

According to the chi-square table the p-value for 1 degree of freedom = <0.001
The underneath chi-square table has an upper row with areas under the curve, a left-end column with degrees of freedom, and a whole lot of chi-square values.

Chi-squared distribution

| | Two-tailed *P*-value | | | |
|---|---|---|---|---|
| *df* | 0.10 | 0.05 | 0.01 | 0.001 |
| 1 | 2.706 | 3.841 | 6.635 | 10.827 |
| 2 | 4.605 | 5.991 | 9.210 | 13.815 |
| 3 | 6.251 | 7.851 | 11.345 | 16.266 |

(continued)

| | Two-tailed *P*-value | | | |
|---|---|---|---|---|
| *df* | 0.10 | 0.05 | 0.01 | 0.001 |
| 4 | 7.779 | 9.488 | 13.277 | 18.466 |
| 5 | 9.236 | 11.070 | 15.086 | 20.515 |
| 6 | 10.645 | 12.592 | 16.812 | 22.457 |
| 7 | 12.017 | 14.067 | 18.475 | 24.321 |
| 8 | 13.362 | 15.507 | 20.090 | 26.124 |
| 9 | 14.684 | 16.919 | 21.666 | 27.877 |
| 10 | 15.987 | 18.307 | 23.209 | 29.588 |
| 11 | 17.275 | 19.675 | 24.725 | 31.264 |
| 12 | 18.549 | 21.026 | 26.217 | 32.909 |
| 13 | 19.812 | 22.362 | 27.688 | 34.527 |
| 14 | 21.064 | 23.685 | 29.141 | 36.124 |
| 15 | 22.307 | 24.996 | 30.578 | 37.698 |
| 16 | 23.542 | 26.296 | 32.000 | 39.252 |
| 17 | 24.769 | 27.587 | 33.409 | 40.791 |
| 18 | 25.989 | 28.869 | 34.805 | 42.312 |
| 19 | 27.204 | 30.144 | 36.191 | 43.819 |
| 20 | 28.412 | 31.410 | 37.566 | 45.314 |
| 21 | 29.615 | 32.671 | 38.932 | 46.796 |
| 22 | 30.813 | 33.924 | 40.289 | 48.268 |
| 23 | 32.007 | 35.172 | 41.638 | 49.728 |
| 24 | 33.196 | 36.415 | 42.980 | 51.179 |
| 25 | 34.382 | 37.652 | 44.314 | 52.619 |
| 26 | 35.536 | 38.885 | 45.642 | 54.051 |
| 27 | 36.741 | 40.113 | 46.963 | 55.475 |
| 28 | 37.916 | 41.337 | 48.278 | 56.892 |
| 29 | 39.087 | 42.557 | 49.588 | 58.301 |
| 30 | 40.256 | 43.773 | 50.892 | 59.702 |
| 40 | 51.805 | 55.758 | 63.691 | 73.403 |
| 50 | 63.167 | 67.505 | 76.154 | 86.660 |
| 60 | 74.397 | 79.082 | 88.379 | 99.608 |
| 70 | 85.527 | 90.531 | 100.43 | 112.32 |
| 80 | 96.578 | 101.88 | 112.33 | 124.84 |
| 90 | 107.57 | 113.15 | 124.12 | 137.21 |
| 100 | 118.50 | 124.34 | 135.81 | 149.45 |

## 5 Data Example, Assessing Heterogeneity

The above example will now be assessed for heterogeneity.

Heterogeneity of this meta-analysis is tested by the fixed effect model.

Heterogeneity chi-square value = 359.516-206.91
= 152.6,
With 9 degrees of freedom the p –value
= <0.001.

Although the meta-analysis shows a significantly lower systolic blood pressure in patients with potassium treatment than those with placebo, this result has a limited meaning, since the studies are significantly heterogeneous. For heterogeneity testing it is tested whether there is a greater inequalities between the results of the separate trials than is compatible with the play of chance. Additional tests for heterogeneity testing are available (Cleophas and Zwinderman, Meta-analysis. In: Statistics Applied to Clinical Studies, Springer New York, 2012, 5th edition, pp 365–388). However, when there is heterogeneity, a careful investigation of its potential cause is often more important than a lot of additional statistical tests.

## 6 Conclusion

Meta-analyses are systematic reviews of multiple published studies with pooled data. Because the separate studies have different sample sizes a weighted average has to be calculated. Heterogeneity in a meta-analysis means that the differences in the results between the studies are larger than could happen by chance. With a significant heterogeneity the meaning of the pooled data is generally little.

Additional tests for heterogeneity testing are available in Statistics Applied to Clinical Studies 5th edition, Chaps 32–34, Springer New York, 2012). However, with significant heterogeneity in a meta-analysis, a careful investigation of its potential cause is more important than lots of statistical tests.

## 7 Note

More background, theoretical and mathematical information of meta-analysis given in Statistics applied to clinical studies 5th edition, Chaps. 32–34, Springer Heidelberg Germany, 2012, from the same authors.

# Chapter 33
# Goodness of Fit Tests for Identifying Nonnormal Data

## 1 General Purpose

Goodness of fit for assessing normal distribution of a data file is an important requirement for normal and t-distributed tests to be sensitive for statistical testing the data. Data files that lack goodness of fit can be analyzed using distribution free methods, like Monte Carlo modeling and neural network modeling (SPSS for Starters, Part 2 from the same authors, Chaps. 18 and 19, Springer New York, 2012, from the same authors).

The chi-square and the Kolmogorov-Smirnov goodness of fit tests are pretty much similar, but one uses the differences between all observed and expected observations, while the other uses the single largest difference between observed and expected observations, and, so, results may not be identical. One test may, however, very well be used as a complementary test or contrast test to the other.

## 2 Schematic Overview of Type of Data File

| Outcome |
|---|
| . |
| . |
| . |
| . |
| . |
| . |
| . |
| . |

(continued)



T.J. Cleophas, A.H. Zwinderman, *Clinical Data Analysis on a Pocket Calculator*,
DOI 10.1007/978-3-319-27104-0_33

| Outcome |
| --- |
| • |
| |
| |
| |

## 3 Primary Scientific Question

Can goodness of fit tests adequately identify nonnormal/non-t-distributed data.

## 4 Chi-Square Goodness of Fit Test, Data Example

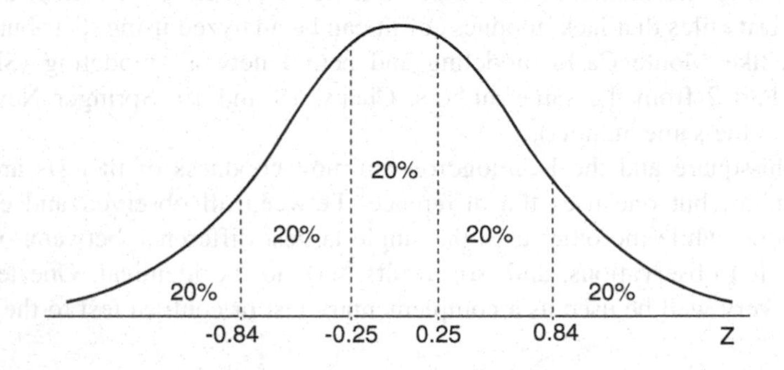


With the help of the t-table the areas under the curve (AUCs) of 5 intervals of the null-hypothesis of a normal frequency distribution can be computed. The cut-off results (z-values) for the 5 intervals with an AUC of 20 % are −0.84, −0.25, 0.25, and 0.84 (AUC = area under the curve). These cut-off results can, subsequently, be used for chi-square goodness of fit testing. A data example is given.

In random not-too-small populations body-weights follow a normal distribution. Is this also true for the body-weights of 50 patients treated with a weight reducing compound?

Individual weight (kgs)

| | | | | | | | | | |
|---|---|---|---|---|---|---|---|---|---|
| 85 | 57 | 60 | 81 | 89 | 63 | 52 | 65 | 77 | 64 |
| 89 | 86 | 90 | 60 | 57 | 61 | 95 | 78 | 66 | 92 |
| 50 | 56 | 95 | 60 | 82 | 55 | 61 | 81 | 61 | 53 |
| 63 | 75 | 50 | 98 | 63 | 77 | 50 | 62 | 79 | 69 |
| 76 | 66 | 97 | 67 | 54 | 93 | 70 | 80 | 67 | 73 |

As the area under the curve (AUC) of a normal distribution curve is divided into 5 equiprobable intervals of 20 % each, we will expect approximately 10 patients per interval. From the data a mean and standard error (SE) of 71 and 15 kg are calculated. In order to compute the numbers of patients of our example in each interval, we will use the underneath equation.

$$z = \text{standardized result} = \frac{\text{unstandardized result} - \text{mean result}}{\text{SE}}$$

For example for the cut-off value of −0.84 the unstandardized result of 58.40 kg can be computed.

−0.84 = (unstandardized result − 71)/15
unstandardized results = (15 × −8.84) + 71 = −58.40 kg.

All of the unstandardized results (kgs) are given underneath:

$-\infty$.... 58.40... 67.25... 74.25... 83.60... $\infty$

As they are equiprobable,

| | | | | | |
|---|---|---|---|---|---|
| As they are equiprobable, we expect per interval: | 10 pts | 10 pts | 10 pts | 10 pts | 10pts |
| We do, however, observe the following numbers: | 10 pts | 16 pts | 3 pts | 10 pts | 11pts |

The chi-square value is calculated according to

$$\sum \frac{(\text{observed number} - \text{expected number})^2}{\text{expected number}} = 8.6$$

This chi-square table is given underneath. It has an upper row with areas under the curve, a left-end column with degrees of freedom, and a whole lot of chi-square values.

Chi-squared distribution.

| | Two-tailed *P*-value | | | |
|---|---|---|---|---|
| *df* | 0.10 | 0.05 | 0.01 | 0.001 |
| 1 | 2.706 | 3.841 | 6.635 | 10.827 |
| 2 | 4.605 | 5.991 | 9.210 | 13.815 |
| 3 | 6.251 | 7.815 | 11.345 | 16.266 |
| 4 | 7.779 | 9.488 | 13.277 | 18.466 |
| 5 | 9.236 | 11.070 | 15.086 | 20.515 |
| 6 | 10.645 | 12.592 | 16.812 | 22.457 |
| 7 | 12.017 | 14.067 | 18.475 | 24.321 |
| 8 | 13.362 | 15.507 | 20.090 | 26.124 |
| 9 | 14.684 | 16.919 | 21.666 | 27.877 |
| 10 | 15.987 | 18.307 | 23.209 | 29.588 |
| 11 | 17.275 | 19.675 | 24.725 | 31.264 |
| 12 | 18.549 | 21.026 | 26.217 | 32.909 |
| 13 | 19.812 | 22.362 | 27.688 | 34.527 |
| 14 | 21.064 | 23.685 | 29.141 | 36.124 |
| 15 | 22.307 | 24.996 | 30.578 | 37.698 |
| 16 | 23.542 | 26.296 | 32.000 | 39.252 |
| 17 | 24.769 | 27.587 | 33.409 | 40.791 |
| 18 | 25.989 | 28.869 | 34.805 | 42.312 |
| 19 | 27.204 | 30.144 | 36.191 | 43.819 |
| 20 | 28.412 | 31.410 | 37.566 | 45.314 |
| 21 | 29.615 | 32.671 | 38.932 | 46.796 |
| 22 | 30.813 | 33.924 | 40.289 | 48.268 |
| 23 | 32.007 | 35.172 | 41.638 | 49.728 |
| 24 | 33.196 | 36.415 | 42.980 | 51.179 |
| 25 | 34.382 | 37.652 | 44.314 | 52.619 |
| 26 | 35.563 | 38.885 | 45.642 | 54.051 |
| 27 | 36.741 | 40.113 | 46.963 | 55.475 |
| 28 | 37.916 | 41.337 | 48.278 | 56.892 |
| 29 | 39.087 | 42.557 | 49.588 | 58.301 |
| 30 | 40.256 | 43.773 | 50.892 | 59.702 |
| 40 | 51.805 | 55.758 | 63.691 | 73.403 |
| 50 | 63.167 | 67.505 | 76.154 | 86.660 |
| 60 | 74.397 | 79.082 | 88.379 | 99.608 |
| 70 | 85.527 | 90.531 | 100.43 | 112.32 |
| 80 | 96.578 | 101.88 | 112.33 | 124.84 |
| 90 | 107.57 | 113.15 | 124.12 | 137.21 |
| 100 | 118.50 | 124.34 | 135.81 | 149.45 |

The chi-square value of 8.6 means that for the given degrees of freedom of 5–1 = 4 (there are 5 different intervals) the null-hypothesis of no-difference-between-observed-and-expected can not be rejected. However, our p-value is <0.10, and, so, there is a trend of a difference. The data may not be entirely normal, as expected. This may be due to lack of randomness.

# 5 Kolmogorov-Smirnov Goodness of Fit Test, Data Example

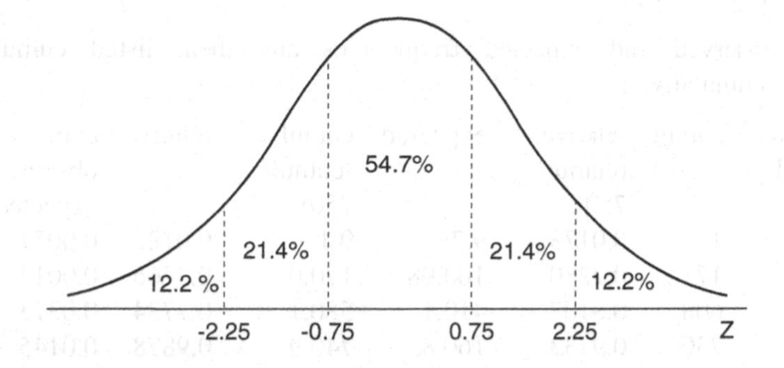


With the help of the t-table the areas under the curve (AUCs) of again 5 intervals of the null-hypothesis of a normal frequency distribution can be computed. The cut-off results (z-values) for the 5 intervals are calculated to be −2.25, −0.75, 0.75, and 2.25. The corresponding AUCs are given in the above graph (AUC = area under the curve).

In random not-too-small populations plasma cholesterol levels follow a normal distribution. Is this also true for the plasma cholesterol levels of the underneath patients treated with a cholesterol reducing compound? A data sample of 750 patients is given.

| Cholesterol (mmol/l) | <4.01 | 4.01–5.87 | 5.87–7.73 | 7.73–9.59 | >9.59 |
|---|---|---|---|---|---|
| Numbers of pts | 13 | 158 | 437 | 122 | 20 |

The cut-off results for the 5 intervals must be standardized to find the expected normal distribution for these data according to

$$z = \text{standardized cut-off result} = \frac{\text{unstandardized result} - \text{mean result}}{\text{SE}}$$

With a calculated mean (SE) of 6.80 (1.24) we must compute the unstandardized results corresponding with the z-values −2.25, −0.75, 0.75 and 2.25. For example, with z = −2.25, the unstandardized z-value is calculated.

$$-2.25 = (\text{unstandardized result} - 6.80)/1.24$$

unstandardized result

$$= (1.24 \times -2.25) + 6.80 = 4.01 \text{ mmol/l}.$$

Similarly all unstandardized z-values are computed.

With 750 cholesterol-values in total the expected frequencies of cholesterol-values in the subsequent intervals are

$12.2 \times 750 = \quad 9.2$
$21.4 \times 750 = \quad 160.8$

54.7 × 750 = 410.1
21.4 × 750 = 160.8
12.2 × 750 = 9.2

The observed and expected frequencies are, then, listed cumulatively (cumul = cumulative):

| Frequency observed | cumul | relative (cumul/ 750) | expected | cumul (cumul/ 750) | relative | cumul observed-expected |
|---|---|---|---|---|---|---|
| 13 | 13 | 0.0173 | 9.2 | 9.1 | 0.0122 | 0.0051 |
| 158 | 171 | 0.2280 | 160.98 | 170.0 | 0.2266 | 0.0014 |
| 437 | 608 | 0.8107 | 410.1 | 580.1 | 0.7734 | 0.0373 |
| 122 | 730 | 0.9733 | 160.8 | 740.9 | 0.9878 | 0.0145 |
| 20 | 750 | 1.000 | 9.2 | 750 | 1.000 | 0.0000 |

According to the Kolmogorov-Smirnov table below, the largest cumulative difference between observed and expected should be smaller than $1.36/\sqrt{n} = 1.36/\sqrt{750} = 0.0497$, while we find 0.0373. This means that these data are well normally distributed.

Level of statistical significance for maximum difference between cumulative observed and expected frequency (n = sample size)

| n | Areas under the curve 0.20 | 0.15 | 0.10 | 0.05 | 0.01 |
|---|---|---|---|---|---|
| 1 | 0.900 | 0.925 | 0.950 | 0.975 | 0.995 |
| 2 | 0.684 | 0.726 | 0.776 | 0.842 | 0.929 |
| 3 | 0.565 | 0.597 | 0.642 | 0.708 | 0.828 |
| 4 | 0.494 | 0.525 | 0.564 | 0.624 | 0.733 |
| 5 | 0.446 | 0.474 | 0.510 | 0.565 | 0.669 |
| 6 | 0.410 | 0.436 | 0.470 | 0.521 | 0.618 |
| 7 | 0.381 | 0.405 | 0.438 | 0.486 | 0.577 |
| 8 | 0.358 | 0.381 | 0.411 | 0.457 | 0.543 |
| 9 | 0.339 | 0.360 | 0.388 | 0.432 | 0.514 |
| 10 | 0.322 | 0.342 | 0.368 | 0.410 | 0.490 |
| 11 | 0.307 | 0.326 | 0.352 | 0.391 | 0.468 |
| 12 | 0.295 | 0.313 | 0.338 | 0.375 | 0.450 |
| 13 | 0.284 | 0.302 | 0.325 | 0.361 | 0.463 |
| 14 | 0.274 | 0.292 | 0.314 | 0.349 | 0.418 |
| 15 | 0.266 | 0.283 | 0.304 | 0.338 | 0.404 |
| 16 | 0.258 | 0.274 | 0.295 | 0.328 | 0.392 |
| 17 | 0.250 | 0.266 | 0.286 | 0.318 | 0.381 |
| 18 | 0.244 | 0.259 | 0.278 | 0.309 | 0.371 |
| 19 | 0.237 | 0.252 | 0.272 | 0.301 | 0.363 |
| 20 | 0.231 | 0.246 | 0.264 | 0.294 | 0.356 |

| | | | | | |
|---|---|---|---|---|---|
| 25 | 0.21 | 0.22 | 0.24 | 0.27 | 0.32 |
| 30 | 0.19 | 0.20 | 0.22 | 0.24 | 0.29 |
| 35 | 0.18 | 0.19 | 0.21 | 0.23 | 0.27 |
| Over 35 | 1.07 √n | 1.14 √n | 1.22 √n | 1.36 √n | 1.63 √n |

# 6 Conclusion

Goodness of fit for assessing normal distribution of a data file is an important requirement for normal and t-distributed tests to be sensitive for statistically testing the data. The chi-square and the Kolmogorov-Smirnov goodness of fit tests are adequate for the purpose, and are pretty much similar, but results need not be identical.

# 7 Note

More background, theoretical and mathematical information of goodness if fit testing is given in Statistics applied to clinical studies 5th edition, Chap. 42, Springer Heidelberg Germany, 2012, from the same authors.

# Chapter 34
# Non-parametric Tests for Three or More Samples (Friedman and Kruskal-Wallis)

## 1 General Purpose

The Friedman test is used for comparing three or more repeated measures that are not normally distributed, and is a non-parametric test, and extension of the Wilcoxon signed rank test (Chap. 6). The Kruskal-Wallis test compares multiple groups that are unpaired and not normally distributed, and is also a non-parametric test, and extension of the Mann–Whitney test (Chap. 7). This chapter assesses both methodologies, that are adequate for non-normal data, but can also be used with normal data.

## 2 Schematic Overview of Type of Data File

| Outcome from treat 1 | From treat 2 | From treat 3 |
|---|---|---|
| . | . | . |
| . | . | . |
| . | . | . |
| . | . | . |
| . | . | . |
| . | . | . |
| . | . | . |
| . | . | . |
| . | . | . |
| | | |
| | | |
| | | |
| | | |
| | | |
| | | |

treat = treatment modality



T.J. Cleophas, A.H. Zwinderman, *Clinical Data Analysis on a Pocket Calculator*,
DOI 10.1007/978-3-319-27104-0_34

## 3 Primary Scientific Question

How do we use the Friedman and Kruskal – Wallis tests for, respectively, testing repeated measures in one group, and single measures in multiple groups, if the assumption of normality is doubtful.

## 4 Friedman Test for Paired Observations

The underneath data are paired comparisons to test effect of 2 dosages of a sleeping drug versus placebo on hours of sleep

| | Hours of sleep | | | | | |
|---|---|---|---|---|---|---|
| Patient | dose 1 | dose 2 | placebo | dose 1 | dose 2 | placebo |
| | (hours) | | | (ranks) | | |
| 1 | 6.1 | 6.8 | 5.2 | 2 | 3 | 1 |
| 2 | 7.0 | 7.0 | 7.9 | 1.5 | 1.5 | 3 |
| 3. | 8.2 | 9.0 | 3.9 | 2 | 3 | 1 |
| 4. | 7.6 | 7.8 | 4.7 | 2 | 3 | 1 |
| 5. | 6.5 | 6.6 | 5.3 | 2 | 3 | 1 |
| 6. | 8.4 | 8.0 | 5.4 | 3 | 2 | 1 |
| 7. | 6.9 | 7.3 | 4.2 | 2 | 3 | 1 |
| 8. | 6.7 | 7.0 | 6.1 | 2 | 3 | 1 |
| 9. | 7.4 | 7.5 | 3.8 | 2 | 3 | 1 |
| 10. | 5.8 | 5.8 | 6.3 | 1.5 | 1.5 | 3 |

The data are ranked for each patient in ascending order of hours of sleep. If the hours are equal, then an average ranknumber is given. Then, for each treatment the squared ranksum is calculated: for dose 1 it equals $(2+1.5+2+2+2+3+2+2+2+1.5)^2=400$, for dose 2 it is 676, for placebo it is 196. The following equation is used:

$$\text{chi-square} = \frac{12}{nk\,(k+1)}\left(\text{ranksum}_{\text{dose1}}{}^2 + \text{ranksum}_{\text{dose2}}{}^2 + \text{ranksum}_{\text{placebo}}{}^2\right) - 3n(k+1),$$

where n = the number of patients and k = the number of treatments.

The chi-square value as calculated is 7.2. The degrees of freedom is $3-1=2$.

The underneath chi-square table has an upper row with areas under the curve, a left-end column with degrees of freedom, and a whole lot of chi-square values. It shows that, for 2 degrees of freedom, a chi-square $> 5.991$ is required to reject the null-hypothesis of no effect at $p < 0.05$.

Chi-squared distribution

| | Two-tailed *P*-value | | | |
|---|---|---|---|---|
| *df* | 0.10 | 0.05 | 0.01 | 0.001 |
| 1 | 2.706 | 3.841 | 6.635 | 10.827 |
| 2 | 4.605 | 5.991 | 9.210 | 13.815 |
| 3 | 6.251 | 7.851 | 11.345 | 16.266 |
| 4 | 7.779 | 9.488 | 13.277 | 18.466 |
| 5 | 9.236 | 11.070 | 15.086 | 20.515 |
| 6 | 10.645 | 12.592 | 16.812 | 22.457 |
| 7 | 12.017 | 14.067 | 18.475 | 24.321 |
| 8 | 13.362 | 15.507 | 20.090 | 26.124 |
| 9 | 14.684 | 16.919 | 21.666 | 27.877 |
| 10 | 15.987 | 18.307 | 23.209 | 29.588 |
| 11 | 17.275 | 19.675 | 24.725 | 31.264 |
| 12 | 18.549 | 21.026 | 26.217 | 32.909 |
| 13 | 19.812 | 22.362 | 27.688 | 34.527 |
| 14 | 21.064 | 23.685 | 29.141 | 36.124 |
| 15 | 22.307 | 24.996 | 30.578 | 37.698 |
| 16 | 23.542 | 26.296 | 32.000 | 39.252 |
| 17 | 24.769 | 27.587 | 33.409 | 40.791 |
| 18 | 25.989 | 28.869 | 34.805 | 42.312 |
| 19 | 27.204 | 30.144 | 36.191 | 43.819 |
| 20 | 28.412 | 31.410 | 37.566 | 45.314 |
| 21 | 29.615 | 32.671 | 38.932 | 46.796 |
| 22 | 30.813 | 33.924 | 40.289 | 48.268 |
| 23 | 32.007 | 35.172 | 41.638 | 49.728 |
| 24 | 33.196 | 36.415 | 42.980 | 51.179 |
| 25 | 34.382 | 37.652 | 44.314 | 52.619 |
| 26 | 35.536 | 38.885 | 45.642 | 54.051 |
| 27 | 36.741 | 40.113 | 46.963 | 55.475 |
| 28 | 37.916 | 41.337 | 48.278 | 56.892 |
| 29 | 39.087 | 42.557 | 49.588 | 58.301 |
| 30 | 40.256 | 43.773 | 50.892 | 59.702 |
| 40 | 51.805 | 55.758 | 63.691 | 73.403 |
| 50 | 63.167 | 67.505 | 76.154 | 86.660 |
| 60 | 74.397 | 79.082 | 88.379 | 99.608 |
| 70 | 85.527 | 90.531 | 100.43 | 112.32 |
| 80 | 96.578 | 101.88 | 112.33 | 124.84 |
| 90 | 107.57 | 113.15 | 124.12 | 137.21 |
| 100 | 118.50 | 124.34 | 135.81 | 149.45 |

Post-hoc subgroups analyses (using Wilcoxon's tests) are required to find out exactly where the difference is situated, between group 1 and 2, between group 1 and 3, or between group 2 and 3 or between two or more groups. The subject of post-hoc testing will be further discussed in the Chaps. 18, 19, and 20.

## 5 Kruskal-Wallis Test for Unpaired Observations

The underneath data show three-samples of patients treated with placebo or 2 different NSAIDs (non steroidal anti-inflammatory drugs). The outcome variable is the fall in plasma globulin concentration (g/l). Group 1 patients are printed in italics, group 2 in normal standard letters, and group 3 in fat prints.

| Globulin concentration (g/l) | ranknumber |
|---|---|
| ***−17*** | ***1*** |
| ***−16*** | ***2*** |
| ***−5*** | ***3*** |
| ***−3*** | ***4*** |
| ***−2*** | ***5*** |
| ***16*** | ***6*** |
| ***18*** | ***7*** |
| 26 | 8 |
| **27** | **9** |
| ***28*** | ***10.5*** |
| **28** | **10.5** |
| 29 | 12 |
| 30 | 14 |
| ***30*** | ***14*** |
| ***30*** | ***14*** |
| 31 | 16 |
| 32 | 17 |
| 33 | 18 |
| **34** | **19** |
| 35 | 20 |
| 36 | 21 |
| 38 | 22.5 |
| **38** | **22.5** |
| **39** | **24.5** |
| **39** | **24.5** |
| **40** | **26** |
| 41 | 27 |
| **42** | **28** |
| **45** | **29.5** |
| **45** | **29.5** |

Three groups of patients with rheumatoid arthritis are treated with a placebo or one of two different NSAIDS. The fall in plasma globulin (g/l) is used to estimate the effect of treatments. First, we will give a ranknumber to every patient dependent on his/her magnitude of fall. If two or three patients have the same fall, they are given an average ranknumber. Then, we calculate the sum of the ranks for the three groups. For group 1 this amounts to $1+2+3+4+5+6+7+10.5+14+14=66.5$, for group 2 to 175.5, group 3 to 488.5. Then we use the equation:

$$\text{chi-square} = \frac{12}{30\,(30-1)}\left(\frac{\text{ranksum}_{\text{group1}}{}^{2}}{10} + \frac{\text{ranksum}_{\text{group2}}{}^{2}}{10} + \frac{\text{ranksum}_{\text{group3}}{}^{2}}{10}\right) - 3(30-1),$$

where the number 30 equals all values, 10 the patient number per group.

The chi-square-value equals 7744.3. This value is very large, indicating that the null-hypothesis of no difference in the data can be rejected. In this example the calculated chi-square value is much larger than the rejection chi-square for (3–1) degrees of freedom and, therefore, we conclude that there is a significant difference between the three treatments at $p < 0.0001$ (see the above chi-square table).

Post-hoc subgroup analyses (using Man-Whitney tests) are required to find out exactly where the difference is situated, between group 1 and 2, between group 1 and 3, or between group 2 and 3 or between two or more groups. The subject of post-hoc testing have been discussed in the Chaps. 18, 19, and 20.

## 6 Conclusion

For the analysis of efficacy data we test null-hypotheses. The t-test (Chaps. 6 and 7) is appropriate for two parallel-groups or two paired samples. Analysis of variance (ANOVA) (Chaps. 19 and 20) is appropriate for analyzing more than two groups / treatments. For data that do not follow a normal frequency distribution, non-parametric tests are available: for paired data the Wilcoxon signed rank (Chap. 6) or Friedman tests (current chapter), for unpaired data the Mann–Whitney test (Chap. 7) or Kruskal-Wallis tests (this chapter) are adequate.

## 7 Note

More background, theoretical and mathematical information of Friedman and Kruskal-Wallis tests are given in Statistics applied to clinical studies 5th edition, Chap. 2, Springer Heidelberg Germany, 2012, from the same authors.

# Part II
# Binary Outcome Data

# Chapter 35
# Data Spread: Standard Deviations, One Sample Z-Test, One Sample Binomial Test

## 1 General Purpose

With continuous outcome data (Chap. 1), the standard deviation is generally used to estimate the spread in a data sample. The standard deviation is, then, used for multiple purposes like null-hypothesis testing and the computation of confidence intervals. With binary outcome data things are different. Instead of a mean value the number of responders is calculated as a "kind of" mean value. In a data sample with binary outcome (yes-no outcome), the spread is estimated with the equation, $\sqrt{(p(1-p)}$, where $p =$ the proportion of responders, otherwise called the (yes-data fraction versus all data). This chapter assesses how these estimators can be used in practice for testing null-hypotheses and confidence intervals of binary outcome data.

## 2 Schematic Overview of Type of Data File

| Outcome binary |
|---|
| . |
| . |
| . |
| . |
| . |
| . |
| . |
| . |
| . |
| |
| |
| |



T.J. Cleophas, A.H. Zwinderman, *Clinical Data Analysis on a Pocket Calculator*,
DOI 10.1007/978-3-319-27104-0_35

## 3 Primary Scientific Question

How do we compute the standard deviation of a binary data set.

## 4 The Computation of the Proportion of Responders and Its Standard Deviation (SD)

Why is SD of the proportion responders = $\sqrt{(p(1-p)}$? The proportion of responders can be looked at as the "mean" of a yes/no data set.

Proportion = mean of yes/no data.
For example, mean of the 6 values [1, 0, 1, 0, 0, 1] is 3/6 yes = 50 % = proportion p.

SD of continuous data = $\sqrt{[\Sigma (a - \bar{a})^2/n}$, where n = sample size.
SD of proportional data = $\sqrt{[p(1 - p)]}$, where p = proportion, e.g., 5/15

What does SD of proportion = $\sqrt{[p(1 - p)]}$ mean in practice?

We assume for example: on average 10/15 in a random population say yes to the question, are you sometimes sleepy through the day. Then, 10/15 saying yes will be encountered most frequently.

The chance to find answer <10/15 or >10/15 gets gradually smaller.

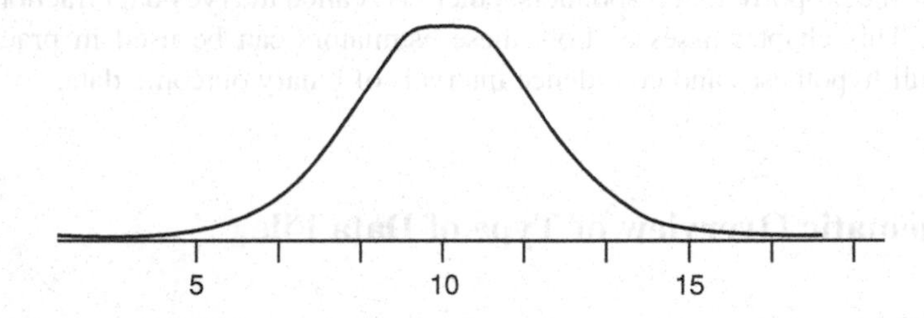


The above graph has on the x-axis the numbers of patients saying yes, and on the y-axis the chance of finding this number. The chance of 8/15 or less is only 15 %, the chance of 7/15 or less only 2.5 %, the chance of 5/15 or less is only 1 %. With many samples the above graph follows a normal frequency distribution, where the equation [$\sqrt{}$ [p(1 – p)] is a very good approximation of its standard deviation. This is how nature works, and it can even be proven to be true with the one sample binomial formula requiring a hypergeometric distribution, but this is beyond the scope of the current work.

## 5 One Sample Z-Test

Out of a sample of 100 patients only 10 patients were yes-responders. And, so, the proportion of yes responders is 10 % = 0.1. For testing, whether this is significantly different from 0 responders, a standard error of the response is required.

The standard error (SE) can be calculated from the standard deviation according to:

SE $= SD/n$, where $n =$ sample size.
SE $= \sqrt{[p(1-p)]}/\sqrt{100}$
$= \sqrt{(0.1 \times 0.9)}/10$
$= 0.03$

the z-value is the test statistic and equals [proportion/(its SE)] $= 0.1/0.03 = 3.33$

The bottom row of the underneath t-table gives p-values from the z-values. With a z-value of 3.33, the p-value, two-tail as usual, should be $<0.001$. This would mean that the 10 % yes responders is significantly better than a zero response would have been.

| *df* | One-Tail = .4<br>Two-Tail = .8 | .25<br>.5 | .1<br>.2 | .05<br>.1 | .025<br>.05 | .01<br>.02 | .005<br>.01 | .0025<br>.005 | .001<br>.002 | .0005<br>.001 |
|---|---|---|---|---|---|---|---|---|---|---|
| 1 | 0.325 | 1.000 | 3.078 | 6.314 | 12.706 | 31.821 | 63.657 | 127.32 | 318.31 | 636.62 |
| 2 | 0.289 | 0.816 | 1.886 | 2.920 | 4.303 | 6.965 | 9.925 | 14.089 | 22.327 | 31.598 |
| 3 | 0.277 | 0.765 | 1.638 | 2.353 | 3.182 | 4.541 | 5.841 | 7.453 | 10.214 | 12.924 |
| 4 | 0.271 | 0.741 | 1.533 | 2.132 | 2.776 | 3.747 | 4.604 | 5.598 | 7.173 | 8.610 |
| 5 | 0.267 | 0.727 | 1.476 | 2.015 | 2.571 | 3.365 | 4.032 | 4.773 | 5.893 | 6.869 |
| 6 | 0.265 | 0.718 | 1.440 | 1.943 | 2.447 | 3.143 | 3.707 | 4.317 | 5.208 | 5.959 |
| 7 | 0.263 | 0.711 | 1.415 | 1.895 | 2.365 | 2.998 | 3.499 | 4.029 | 4.785 | 5.408 |
| 8 | 0.262 | 0.706 | 1.397 | 1.860 | 2.306 | 2.896 | 3.355 | 3.833 | 4.501 | 5.041 |
| 9 | 0.261 | 0.703 | 1.383 | 1.833 | 2.262 | 2.821 | 3.250 | 3.690 | 4.297 | 4.781 |
| 10 | 0.260 | 0.700 | 1.372 | 1.812 | 2.228 | 2.764 | 3.169 | 3.581 | 4.144 | 4.587 |
| 11 | 0.260 | 0.697 | 1.363 | 1.796 | 2.201 | 2.718 | 3.106 | 3.497 | 4.025 | 4.437 |
| 12 | 0.259 | 0.695 | 1.356 | 1.782 | 2.179 | 2.681 | 3.055 | 3.428 | 3.930 | 4.318 |
| 13 | 0.259 | 0.694 | 1.350 | 1.771 | 2.160 | 2.650 | 3.012 | 3.372 | 3.852 | 4.221 |
| 14 | 0.258 | 0.692 | 1.345 | 1.761 | 2.145 | 2.624 | 2.977 | 3.326 | 3.787 | 4.140 |
| 15 | 0.258 | 0.691 | 1.341 | 1.753 | 2.131 | 2.602 | 2.947 | 3.286 | 3.733 | 4.073 |
| 16 | 0.258 | 0.690 | 1.337 | 1.746 | 2.120 | 2.583 | 2.921 | 3.252 | 3.686 | 4.015 |
| 17 | 0.257 | 0.689 | 1.333 | 1.740 | 2.110 | 2.567 | 2.898 | 3.222 | 3.646 | 3.965 |
| 18 | 0.257 | 0.688 | 1.330 | 1.734 | 2.101 | 2.552 | 2.878 | 3.197 | 3.610 | 3.922 |
| 19 | 0.257 | 0.688 | 1.328 | 1.729 | 2.093 | 2.539 | 2.861 | 3.174 | 3.579 | 3.883 |
| 20 | 0.257 | 0.687 | 1.325 | 1.725 | 2.086 | 2.528 | 2.845 | 3.153 | 3.552 | 3.850 |
| 21 | 0.257 | 0.686 | 1.323 | 1.721 | 2.080 | 2.518 | 2.831 | 3.135 | 3.527 | 3.819 |
| 22 | 0.256 | 0.686 | 1.321 | 1.717 | 2.074 | 2.508 | 2.819 | 3.119 | 3.505 | 3.792 |
| 23 | 0.256 | 0.685 | 1.319 | 1.714 | 2.069 | 2.500 | 2.807 | 3.104 | 3.485 | 3.767 |
| 24 | 0.256 | 0.685 | 1.318 | 1.711 | 2.064 | 2.492 | 2.797 | 3.091 | 3.467 | 3.745 |
| 25 | 0.256 | 0.684 | 1.316 | 1.708 | 2.060 | 2.485 | 2.787 | 3.078 | 3.450 | 3.725 |
| 26 | 0.256 | 0.684 | 1.315 | 1.706 | 2.056 | 2.479 | 2.779 | 3.067 | 3.435 | 3.707 |
| 27 | 0.256 | 0.684 | 1.314 | 1.703 | 2.052 | 2.473 | 2.771 | 3.057 | 3.421 | 3.690 |
| 28 | 0.256 | 0.683 | 1.313 | 1.701 | 2.048 | 2.467 | 2.763 | 3.047 | 3.408 | 3.674 |
| 29 | 0.256 | 0.683 | 1.311 | 1.699 | 2.045 | 2.462 | 2.756 | 3.038 | 3.396 | 3.659 |
| 30 | 0.256 | 0.683 | 1.310 | 1.697 | 2.042 | 2.457 | 2.750 | 3.030 | 3.385 | 3.646 |
| 40 | 0.255 | 0.681 | 1.303 | 1.684 | 2.021 | 2.423 | 2.704 | 2.971 | 3.307 | 3.551 |
| 60 | 0.254 | 0.679 | 1.296 | 1.671 | 2.000 | 2.390 | 2.660 | 2.915 | 3.232 | 3.460 |
| 120 | 0.254 | 0.677 | 1.289 | 1.658 | 1.980 | 2.358 | 2.617 | 2.860 | 3.160 | 3.373 |
| ∞ | 0.253 | 0.674 | 1.282 | 1.645 | 1.960 | 2.326 | 2.576 | 2.807 | 3.090 | 3.291 |

The t-table has a left-end column giving degrees of freedom (≈ sample sizes), and two top rows with p-values (areas under the curve = p-values), one-tail meaning that only one end of the curve, two-tail meaning that both ends are assessed simultaneously. The t-table is, furthermore, full of t-values, that, with ∞ degrees of freedom, are equal to z-values. The z-values and t-values are to be understood as mean results of studies, but not expressed in mmol/l, kilograms, or proportions of responders, but in so-called SEM-units (Standard error of the mean units), that are obtained by dividing your mean result by its own standard error. For continuous outcome data, with many degrees of freedom (large samples) the curve will be a little bit narrower, and more in agreement with nature. For binary outcome data, nature has determined that the curves will always be as narrow as can be, according to the row at the bottom.

# 6 Computing the 95 % Confidence Intervals of a Data Set

The example is taken from the Chap. 13. What is the standard error (SE) of a study with events in 10 % of the patients, and a sample size of 100 (n). Ten percent events means a proportion of events of 0.1. The standard deviation (SD) of this proportion is defined by the equation

$\sqrt{}$ [proportion × (1 – proportion)] $= \sqrt{(0.1 \times 0.9)} = \sqrt{0.09} = 0.3$,

the standard error = standard deviation / $\sqrt{n}$,
$= 0.3 / 10 = 0.03$,

the 95 % confidence interval is given by

proportion given ± 1.960 × 0.03 $= 0.1 \pm 1.960 \times 0.03$,
$= 0.1 \pm 0.06$,
= between 0.04 and 0.16.

The 95 % confidence intervals can be used for multiple purposes, for example for noninferiority testing (Chap. 15), and equivalence testing (Chap. 14).

# 7 Conclusion

With binary outcome data, instead of a mean value the number of responders is calculated as a "kind of" mean value. The spread is estimated with the equation, √(p (1 – p), where p = the proportion of responders, otherwise called the (yes-data fraction from all data), in a data sample. This chapter assesses how these estimators can be used in practice for testing null-hypotheses and confidence intervals of binary data.

# 8 Note

More background, theoretical and mathematical information of standard deviations of binary data is given in Statistics applied to clinical studies 5th edition, Chap. 3, Springer Heidelberg Germany, 2012, from the same authors.

# Chapter 36
# Z-Test for Cross-Tabs

## 1 General Purpose

If, as shown in the previous chapter, multiple similar a selective samples of binary data have a normal frequency distribution, then t-tests, like the ones for continuous data (Chap. 7), should be OK for their analysis. Consequently, if we compare two binary samples, for example the outcomes of two treatment modalities, then a two sample t-test should be OK for statistical testing. The two sample t-test for binary outcomes is called the two sample z-test. This chapter shows how it works.

## 2 Schematic Overview of Type of Data File

| Outcome (binary) | predictor (binary) |
|---|---|
| . | |
| . | |
| . | |
| . | |
| . | |
| . | |
| . | |
| . | |
| . | |
| | |
| | |
| | |
| | |
| | |



T.J. Cleophas, A.H. Zwinderman, *Clinical Data Analysis on a Pocket Calculator*,
DOI 10.1007/978-3-319-27104-0_36

## 3 Primary Scientific Question

Is the two sample z-test adequate for comparing the differences in numbers of responders to two different treatment modalities.

## 4 Data Example, Two Group Z-Test

Two groups of patients are assessed for being sleepy through the day. We wish to estimate whether group 1 is more sleepy than group 2. The underneath cross-tab gives the data.

| | Sleepiness | no sleepiness |
|---|---|---|
| Treatment 1 (group 1) | 5 (a) | 10 (b) |
| Treatment 2 (group 2) | 9 (c) | 6 (d) |

$$z = \frac{\text{difference between proportions of sleepers per group (d)}}{\text{pooled standard error difference}}$$

$$z = \frac{d}{\text{pooled SE}} = \frac{(9/15 - 5/15)}{\sqrt{(SE_1^2 + SE_2^2)}}$$

$$SE_1 \text{ ( or } SEM_1 \text{ )} = \sqrt{\frac{p_1(1 - p_1)}{n_1}} \quad \text{where } p_1 = 5/15 \text{ etc........,}$$

$z = 1.45$, not statistically significant from zero, because for a two- tail p – value <0.05 a z-value of at least 1.96 is required. This means that no significant difference between the two groups is observed. The p-value of the z-test can be obtained by using the bottom row of the underneath t-table.

| df | One-Tail = .4<br>Two-Tail = .8 | .25<br>.5 | .1<br>.2 | .05<br>.1 | .025<br>.05 | .01<br>.02 | .005<br>.01 | .0025<br>.005 | .001<br>.002 | .0005<br>.001 |
|---|---|---|---|---|---|---|---|---|---|---|
| 1 | 0.325 | 1.000 | 3.078 | 6.314 | 12.706 | 31.821 | 63.657 | 127.32 | 318.31 | 636.62 |
| 2 | 0.289 | 0.816 | 1.886 | 2.920 | 4.303 | 6.965 | 9.925 | 14.089 | 22.327 | 31.598 |
| 3 | 0.277 | 0.765 | 1.638 | 2.353 | 3.182 | 4.541 | 5.841 | 7.453 | 10.214 | 12.924 |
| 4 | 0.271 | 0.741 | 1.533 | 2.132 | 2.776 | 3.747 | 4.604 | 5.598 | 7.173 | 8.610 |
| 5 | 0.267 | 0.727 | 1.476 | 2.015 | 2.571 | 3.365 | 4.032 | 4.773 | 5.893 | 6.869 |
| 6 | 0.265 | 0.718 | 1.440 | 1.943 | 2.447 | 3.143 | 3.707 | 4.317 | 5.208 | 5.959 |
| 7 | 0.263 | 0.711 | 1.415 | 1.895 | 2.365 | 2.998 | 3.499 | 4.029 | 4.785 | 5.408 |
| 8 | 0.262 | 0.706 | 1.397 | 1.860 | 2.306 | 2.896 | 3.355 | 3.833 | 4.501 | 5.041 |
| 9 | 0.261 | 0.703 | 1.383 | 1.833 | 2.262 | 2.821 | 3.250 | 3.690 | 4.297 | 4.781 |
| 10 | 0.260 | 0.700 | 1.372 | 1.812 | 2.228 | 2.764 | 3.169 | 3.581 | 4.144 | 4.587 |
| 11 | 0.260 | 0.697 | 1.363 | 1.796 | 2.201 | 2.718 | 3.106 | 3.497 | 4.025 | 4.437 |
| 12 | 0.259 | 0.695 | 1.356 | 1.782 | 2.179 | 2.681 | 3.055 | 3.428 | 3.930 | 4.318 |
| 13 | 0.259 | 0.694 | 1.350 | 1.771 | 2.160 | 2.650 | 3.012 | 3.372 | 3.852 | 4.221 |
| 14 | 0.258 | 0.692 | 1.345 | 1.761 | 2.145 | 2.624 | 2.977 | 3.326 | 3.787 | 4.140 |
| 15 | 0.258 | 0.691 | 1.341 | 1.753 | 2.131 | 2.602 | 2.947 | 3.286 | 3.733 | 4.073 |
| 16 | 0.258 | 0.690 | 1.337 | 1.746 | 2.120 | 2.583 | 2.921 | 3.252 | 3.686 | 4.015 |
| 17 | 0.257 | 0.689 | 1.333 | 1.740 | 2.110 | 2.567 | 2.898 | 3.222 | 3.646 | 3.965 |
| 18 | 0.257 | 0.688 | 1.330 | 1.734 | 2.101 | 2.552 | 2.878 | 3.197 | 3.610 | 3.922 |
| 19 | 0.257 | 0.688 | 1.328 | 1.729 | 2.093 | 2.539 | 2.861 | 3.174 | 3.579 | 3.883 |
| 20 | 0.257 | 0.687 | 1.325 | 1.725 | 2.086 | 2.528 | 2.845 | 3.153 | 3.552 | 3.850 |
| 21 | 0.257 | 0.686 | 1.323 | 1.721 | 2.080 | 2.518 | 2.831 | 3.135 | 3.527 | 3.819 |
| 22 | 0.256 | 0.686 | 1.321 | 1.717 | 2.074 | 2.508 | 2.819 | 3.119 | 3.505 | 3.792 |
| 23 | 0.256 | 0.685 | 1.319 | 1.714 | 2.069 | 2.500 | 2.807 | 3.104 | 3.485 | 3.767 |
| 24 | 0.256 | 0.685 | 1.318 | 1.711 | 2.064 | 2.492 | 2.797 | 3.091 | 3.467 | 3.745 |
| 25 | 0.256 | 0.684 | 1.316 | 1.708 | 2.060 | 2.485 | 2.787 | 3.078 | 3.450 | 3.725 |
| 26 | 0.256 | 0.684 | 1.315 | 1.706 | 2.056 | 2.479 | 2.779 | 3.067 | 3.435 | 3.707 |
| 27 | 0.256 | 0.684 | 1.314 | 1.703 | 2.052 | 2.473 | 2.771 | 3.057 | 3.421 | 3.690 |
| 28 | 0.256 | 0.683 | 1.313 | 1.701 | 2.048 | 2.467 | 2.763 | 3.047 | 3.408 | 3.674 |
| 29 | 0.256 | 0.683 | 1.311 | 1.699 | 2.045 | 2.462 | 2.756 | 3.038 | 3.396 | 3.659 |
| 30 | 0.256 | 0.683 | 1.310 | 1.697 | 2.042 | 2.457 | 2.750 | 3.030 | 3.385 | 3.646 |
| 40 | 0.255 | 0.681 | 1.303 | 1.684 | 2.021 | 2.423 | 2.704 | 2.971 | 3.307 | 3.551 |
| 60 | 0.254 | 0.679 | 1.296 | 1.671 | 2.000 | 2.390 | 2.660 | 2.915 | 3.232 | 3.460 |
| 120 | 0.254 | 0.677 | 1.289 | 1.658 | 1.980 | 2.358 | 2.617 | 2.860 | 3.160 | 3.373 |
| ∞ | 0.253 | 0.674 | 1.282 | 1.645 | 1.960 | 2.326 | 2.576 | 2.807 | 3.090 | 3.291 |

The t-table has a left-end column giving degrees of freedom (≈ sample sizes), and two top rows with p-values (areas under the curve = p – values), one-tail meaning that only one end of the curve, two-tail meaning that both ends are assessed simultaneously. The t-table is, furthermore, full of t-values, that, with ∞ degrees of freedom, are equal to z-values. The z-values and t-values are to be understood as mean results of studies, but not expressed in mmol/l, kilograms, or proportions of responders, but in so-called SEM-units (Standard error of the mean units), that are obtained by dividing your mean result by its own standard error. For continuous outcome data, with many degrees of freedom (large samples) the curve will be a little bit narrower, and more in agreement with nature. For binary outcome data, nature has determined that the curves will always be as narrow as can be, according to the row at the bottom.

## 5 Single Group Z-Test

A single group z-test is also possible (see also Chap. 35). For example, in 10 patients we have 4 responders. We question whether 4 responders is significantly more than 0 responders.

$$
\begin{aligned}
z &= \text{proportion/(its SE)} \\
SE &= \surd\,[(4/10 \times (1 - 4/10))/n] \\
&= \surd\,(0.24/10) \\
z &= 0.4/\surd\,(0.24/10) \\
z &= 0.4/0.1549 \\
z &= 2.582
\end{aligned}
$$

According to the bottom row of the t-table the p-value is $<0.01$. A proportion of 0.4 is, thus, significantly larger than a proportion of 0.0.

## 6 Conclusion

As multiple similar a selective samples of binary data have a normal frequency distribution, the t-tests like the ones for continuous data (Chap. 7) are OK for their analysis. Consequently, for comparing two binary samples, for example the outcomes of two treatment modalities, two sample t-test is OK for statistical testing. The two sample t-test for binary outcomes is called the two sample z-test. This chapter shows how it works.

## 7 Note

More background, theoretical and mathematical information is given in the Statistics applied to clinical studies 5th edition, Springer Heidelberg Germany, Chap. 3, 2012, from the same authors.

# Chapter 37
# Phi Tests for Nominal Data

## 1 General Purpose

Nominal data are the simplest type of data. Unlike ordinal data (Chap. 9) and continuous data (Chaps. 1, 2, 3, 4, 5, 6, 7, 8, 9, 10, 11, 12, 13, 14, 15, 16, 17, 18, 19, 20, 21, 22, 23, 24, 25, 26, 27, 28, 29, 30, 31, 32, 33, and 34), they are assumed not to have a stepping function. Examples are genders, age classes, family names. Of nominal data the simplest versions are the bifurcated data (binary data, dichotomous data, yes no data). Chi-square tests can be used for analysis, but they do not provide levels of association, which may be clinically rather relevant. As an example, males and females may be assessed for successful exams.

| | success | yes | no |
|---|---|---|---|
| males | | 50(a) | 25(b) |
| females | | 50(c) | 25(d) |

The value of [(a × d) – (b × c)] can be used to estimate the level of association. In the above example the level of association is 0. The gender does not give the faintest prediction of the chance of a successful exam.

| | success | yes | no |
|---|---|---|---|
| males | | 50(a) | 0(b) |
| females | | 0(c) | 50(d) |

In the above example the level of association equals 1 (100 %). The outcome predicts the chance of a successful exam with 100 % certainty. Phi's, otherwise called Cramer's V's, are used to calculate the precise level of association, being between −1 and 1, and can be easily tested for statistical significance with the help of a chi-square test.



T.J. Cleophas, A.H. Zwinderman, *Clinical Data Analysis on a Pocket Calculator*,
DOI 10.1007/978-3-319-27104-0_37

## 2 Schematic Overview of Type of Data File

| Predictor (binary) | outcome (binary) |
|---|---|
| . | . |
| . | . |
| . | . |
| . | . |
| . | . |
| . | . |
| . | . |
| . | . |
| . | . |
| | |
| | |

## 3 Primary Scientific Question

In a data sample of binary variables, what is the level of association between a binary predictor and another binary outcome variable.

## 4 Data Example

In a hospital many patients tend to fall out of bed. We wish to find out whether one department performs better than the other.

| | fall out of bed | yes | no |
|---|---|---|---|
| department 1 | | 15(a) | 20(b) |
| department 2 | | 15(c) | 5(d) |

$$\text{phi} = (ad-bc)/\surd\,(a+b)(c+d)(a+c)(b+d) =$$
$$= (300\text{–}75)/\surd\,(35 \times 20 \times 30 \times 25) =$$
$$= 0.31$$

We can predict the chance of falling out of bed from the department by 0.31 = 31 %. Knowing the department, you will be 31 % certain about the chance of falling out of bed. Is this level of association statistically significant? this means it 31 % significantly more than 0 %. This of course depends on the magnitude of the samples.

With very small samples it is hard to obtain statistical significance. The underneath chi-square equation is adequate for statistical testing.

$$\begin{aligned} \text{chi-square} &= \text{phi}^2 \times \text{n, where n} = (\text{a} + \text{b} + \text{c} + \text{d}) \\ &= 0.31^2 \times 55 \\ &= 5.29 \end{aligned}$$

The underneath chi-square table shows areas under the curve in the top row, and (df) degrees of freedom in the left-end column, and furthermore plenty of chi-square values. The table shows, that a chi-square value of 5.29 with 1 df (degree of freedom), $((2–1) \times (2–1) = 1)$ is between 3.841 and 6.635. This means that the p-value is between 0.05 and 0.01, and, thus, $< 0.05$. The association is significantly better than an association of zero, no association at all, at $p < 0.05$.

Chi-squared distribution

| | Two-tailed *P*-value | | | |
|---|---|---|---|---|
| *df* | 0.10 | 0.05 | 0.01 | 0.001 |
| 1 | 2.706 | 3.841 | 6.635 | 10.827 |
| 2 | 4.605 | 5.991 | 9.210 | 13.815 |
| 3 | 6.251 | 7.851 | 11.345 | 16.266 |
| 4 | 7.779 | 9.488 | 13.277 | 18.466 |
| 5 | 9.236 | 11.070 | 15.086 | 20.515 |
| 6 | 10.645 | 12.592 | 16.812 | 22.457 |
| 7 | 12.017 | 14.067 | 18.475 | 24.321 |
| 8 | 13.362 | 15.507 | 20.090 | 26.124 |
| 9 | 14.684 | 16.919 | 21.666 | 27.877 |
| 10 | 15.987 | 18.307 | 23.209 | 29.588 |
| 11 | 17.275 | 19.675 | 24.725 | 31.264 |
| 12 | 18.549 | 21.026 | 26.217 | 32.909 |
| 13 | 19.812 | 22.362 | 27.688 | 34.527 |
| 14 | 21.064 | 23.685 | 29.141 | 36.124 |
| 15 | 22.307 | 24.996 | 30.578 | 37.698 |
| 16 | 23.542 | 26.296 | 32.000 | 39.252 |
| 17 | 24.769 | 27.587 | 33.409 | 40.791 |
| 18 | 25.989 | 28.869 | 34.805 | 42.312 |
| 19 | 27.204 | 30.144 | 36.191 | 43.819 |
| 20 | 28.412 | 31.410 | 37.566 | 45.314 |
| 21 | 29.615 | 32.671 | 38.932 | 46.796 |
| 22 | 30.813 | 33.924 | 40.289 | 48.268 |
| 23 | 32.007 | 35.172 | 41.638 | 49.728 |
| 24 | 33.196 | 36.415 | 42.980 | 51.179 |
| 25 | 34.382 | 37.652 | 44.314 | 52.619 |
| 26 | 35.536 | 38.885 | 45.642 | 54.051 |

(continued)

| Chi-squared distribution | | | | |
|---|---|---|---|---|
| | Two-tailed *P*-value | | | |
| *df* | 0.10 | 0.05 | 0.01 | 0.001 |
| 27 | 36.741 | 40.113 | 46.963 | 55.475 |
| 28 | 37.916 | 41.337 | 48.278 | 56.892 |
| 29 | 39.087 | 42.557 | 49.588 | 58.301 |
| 30 | 40.256 | 43.773 | 50.892 | 59.702 |
| 40 | 51.805 | 55.758 | 63.691 | 73.403 |
| 50 | 63.167 | 67.505 | 76.154 | 86.660 |
| 60 | 74.397 | 79.082 | 88.379 | 99.608 |
| 70 | 85.527 | 90.531 | 100.43 | 112.32 |
| 80 | 96.578 | 101.88 | 112.33 | 124.84 |
| 90 | 107.57 | 113.15 | 124.12 | 137.21 |
| 100 | 118.50 | 124.34 | 135.81 | 149.45 |

## 5 Conclusion

Nominal data are the simplest type of data. Unlike ordinal data (Chap. 9) and continuous data (Chaps. 1, 2, 3, 4, 5, 6, 7, 8, 9, 10, 11, 12, 13, 14, 15, 16, 17, 18, 19, 20, 21, 22, 23, 24, 25, 26, 27, 28, 29, 30, 31, 32, 33, and 34), they have no stepping function. Of nominal data the simplest versions are the bifurcated data (binary data). Chi-square tests can be used for analysis (Chap. 38), but they do not provide levels of association, which may be clinically rather relevant. Phi's, otherwise called Cramer's V's, are used to calculate the precise level of association, and can be additionally tested for statistical significance with the help of a chi-square test.

## 6 Note

More background, theoretical and mathematical information of binary data is given in Statistics applied to clinical studies 5th edition, Chap. 3, Springer Heidelberg Germany, 2012, from the same authors.

# Chapter 38
# Chi-square Tests

## 1 General Purpose

Z-tests (Chap. 36) are OK for comparing the effects of two treatment modalities on numbers of responders to treatment, however, pretty laborious. Phi tests (Chap. 37) provide levels of association (or interaction) between a treatment modality and the number of responders, but no p-values. Often, we wish to know whether our result or the difference between two results are significantly different from zero. Chi-square tests make use of chi-square distributions (squared normal distributions), and produce p-values for the purpose, and these p-values are pretty much similar to those of the z-tests, but they can be obtained more easily.

## 2 Schematic Overview of Type of Data File

| Predictor (binary) | Outcome (binary) |
|---|---|
| . | . |
| . | . |
| . | . |
| . | . |
| . | . |
| . | . |
| . | . |
| . | . |
| . | . |



T.J. Cleophas, A.H. Zwinderman, *Clinical Data Analysis on a Pocket Calculator*,
DOI 10.1007/978-3-319-27104-0_38

## 3 Primary Scientific Question

In a $2 \times 2$ interaction cross-tab of binary data, like a study of two treatment modalities and the chance of responding or not in the outcome, is there a significant difference between the two treatment modalities and the outcome.

## 4 Data Example I

The underneath table shows two groups assessed for suffering from sleepiness through the day after being treated with two different anti-sleepiness medications. We wish to know whether there is a significant difference between the proportions of subjects being sleepy.

| | Sleepiness | no sleepiness | |
|---|---|---|---|
| Group 1 | 5(a) | 10(b) | 15(a+b) |
| Group 2 | 9(c) | 6(d) | 15(c+d) |
| | 14(a+c) | 16(b+d) | 30(a+b+c+d) |

The chi-square pocket calculator method is used for testing these data.

$$\chi^2 = \frac{(ad - bc)^2(a + b + c + d)}{(a + b)\,(c + d)\,(b + d)\,(a + c)} = \frac{(30 - 90)^2(30)}{15 \times 15 \times 16 \times 14} = \frac{3600 \times 30}{15 \times 15 \times 16 \times 14}$$

$$= \frac{108.000}{50.400} = 2.143$$

The chi-square value equals 2.143. The underneath chi-square table can tell you, whether or not the difference between the groups is statistically significant.

Chi-squared distribution

| | Two-tailed *P*-value | | | |
|---|---|---|---|---|
| *df* | 0.10 | 0.05 | 0.01 | 0.001 |
| 1 | 2.706 | 3.841 | 6.635 | 10.827 |
| 2 | 4.605 | 5.991 | 9.210 | 13.815 |
| 3 | 6.251 | 7.851 | 11.345 | 16.266 |
| 4 | 7.779 | 9.488 | 13.277 | 18.466 |
| 5 | 9.236 | 11.070 | 15.086 | 20.515 |
| 6 | 10.645 | 12.592 | 16.812 | 22.457 |
| 7 | 12.017 | 14.067 | 18.475 | 24.321 |
| 8 | 13.362 | 15.507 | 20.090 | 26.124 |
| 9 | 14.684 | 16.919 | 21.666 | 27.877 |
| 10 | 15.987 | 18.307 | 23.209 | 29.588 |
| 11 | 17.275 | 19.675 | 24.725 | 31.264 |
| 12 | 18.549 | 21.026 | 26.217 | 32.909 |
| 13 | 19.812 | 22.362 | 27.688 | 34.527 |
| 14 | 21.064 | 23.685 | 29.141 | 36.124 |
| 15 | 22.307 | 24.996 | 30.578 | 37.698 |
| 16 | 23.542 | 26.296 | 32.000 | 39.252 |
| 17 | 24.769 | 27.587 | 33.409 | 40.791 |
| 18 | 25.989 | 28.869 | 34.805 | 42.312 |
| 19 | 27.204 | 30.144 | 36.191 | 43.819 |
| 20 | 28.412 | 31.410 | 37.566 | 45.314 |
| 21 | 29.615 | 32.671 | 38.932 | 46.796 |
| 22 | 30.813 | 33.924 | 40.289 | 48.268 |
| 23 | 32.007 | 35.172 | 41.638 | 49.728 |
| 24 | 33.196 | 36.415 | 42.980 | 51.179 |
| 25 | 34.382 | 37.652 | 44.314 | 52.619 |
| 26 | 35.536 | 38.885 | 45.642 | 54.051 |
| 27 | 36.741 | 40.113 | 46.963 | 55.475 |
| 28 | 37.916 | 41.337 | 48.278 | 56.892 |
| 29 | 39.087 | 42.557 | 49.588 | 58.301 |
| 30 | 40.256 | 43.773 | 50.892 | 59.702 |
| 40 | 51.805 | 55.758 | 63.691 | 73.403 |
| 50 | 63.167 | 67.505 | 76.154 | 86.660 |
| 60 | 74.397 | 79.082 | 88.379 | 99.608 |
| 70 | 85.527 | 90.531 | 100.43 | 112.32 |
| 80 | 96.578 | 101.88 | 112.33 | 124.84 |
| 90 | 107.57 | 113.15 | 124.12 | 137.21 |
| 100 | 118.50 | 124.34 | 135.81 | 149.45 |

In summary, the chi-square table has an upper row with areas under the curve (p-values), a left-end column with degrees of freedom, and a whole lot of chi-square values.

The chi-square table has columns and rows. The upper row are the p-values. The left-end column are the degrees of freedom (df), which are, here, largely in agreement with the numbers of cells in a cross-tab. The simplest cross-tab has 4 cells, which means $2 \times 2 = 4$ cells. The table has been constructed such, that we have here $(2–1) \times (2–1) = 1$ degree of freedom. Look at the row with 1 degree of freedom: a chi-square value of 2.143 is left from 2.706. Now look from here right up at the upper row. The corresponding p-value is larger than 0.1 (10 %). There is, thus, no significant difference in sleepiness between the two groups. The small difference observed is due to the play of chance.

## 5 Data Example II

Two partnerships of internists have the intention to associate. However, in one of the two a considerable number of internists has suffered from a burn-out.

| | burn out | no burn out | |
|---|---|---|---|
| partnership 1 | 3(a) | 7(b) | 10(a + b) |
| partnership 2 | 0(c) | 10(d) | 10(c + d) |
| | 3(a + c) | 17(b + d) | 20(a + b + c + d) |

$$\chi^2 = \frac{(ad - bc)^2 (a + b + c + d)}{(a + b)\ (c + d)\ (b + d)\ (a + c)} = \frac{(30 - 0)\ \ 2(20)}{10 \times 10 \times 17 \times 3} = \frac{900 \times 20}{\ldots\ldots\ldots} = 3.529$$

According to the chi-square table of the previous page a p-value is found of <0.10. This means that no significant difference is found, but a p-value between 0.05 and 0.10 is looked upon as a trend to significance. The difference may be due to some avoidable or unavoidable cause. We should add, here, that values in a cell lower than 5 is considered slightly inappropriate according to some, and another test like the log likelihood ratio test (Chap. 46) is more safe.

## 6 Example for Practicing 1

| Example 2 × 2 table | events | no events | |
|---|---|---|---|
| group 1 | 15(a) | 20(b) | 35(a + b) |
| group 2 | 15(c) | 5(d) | 20(c + d) |
| | 30(a + c) | 25(b + d) | 55(a + b + c + d) |

Pocket calculator

$$\frac{(ad - bc)^2 (a + b + c + d)}{(a + b)(c + d)(b + d)(a + c)} = \quad p=\ldots$$

## 7 Example for Practicing 2

| Another example 2 × 2 table | events | no events | |
|---|---|---|---|
| group 1 | 16(a) | 26(b) | 42(a + b) |
| group 2 | 5(c) | 30(d) | 35(c + d) |
| | 21(a + c) | 56(b + d) | 77(a + b + c + d) |

Pocket calculator

$$\frac{(ad - bc)^2 (a + b + c + d)}{(a + b)(c + d)(b + d)(a + c)} = \quad p=\ldots$$

## 8 Conclusion

Phi tests (previous chapter) provide levels of association (or interaction) between a treatment modality and the number of responders, but no p-values. Often, we wish to know whether our result is significantly different from a zero-result. Chi-square tests make use of the chi-square distribution (squared normal distribution), and produce p-values for the purpose. A p-value $< 0.05$ means that our result is better than a zero-result with over 95 % certainty, or with less than 5 % chance of a type I error (the chance of finding a difference where there is none). A pleasant thing about the pocket-calculator-chi-square-test is, that it can be readily performed during working hours by physicians, for the purpose of solving the scientific questions of their daily life (Chaps. 58, and 59). This means no need for starting a major statistical software program like SPSS, SAS, or R.

## 9 Note

More background, theoretical and mathematical information of chi-square tests is given in Statistics applied to clinical studies 5th edition, Chap. 3, Springer Heidelberg Germany, 2012, from the same authors.

# Chapter 39
# Fisher Exact Tests Convenient for Small Samples

## 1 General Purpose

Fisher-exact test is used as a test for the analysis of cross tabs, and as a contrast test to the chi-square test (Chap. 38), and the z-test (Chap. 36), and, also, as a binary outcome test for small samples, e.g., samples of $n < 100$. It, essentially, makes use of faculties expressed as the sign "!": e.g.,
[5!] indicating $5 \times 4 \times 3 \times 2 \times 1$.

In the past, it was rather laborious with large data, but nowadays any pocket calculator calculates largest faculties within seconds. E.g., using the Scientific Calculator from the Chap. 1, you press 6 and then the 2ndF button, and the CE button: In the display is 720, which is equal to

$6! = 6 \times 5 \times 4 \times 3 \times 2 \times 1 = 720.$

This chapter assesses the performance of the faculty-based Fisher exact test as compared to the traditional distribution-based methods.

## 2 Schematic Overview of Type of Data File

| Outcome binary | Predictor binary |
|---|---|
| . | . |
| . | . |
| . | . |
| . | . |
| . | . |
| . | . |

(continued)



T.J. Cleophas, A.H. Zwinderman, *Clinical Data Analysis on a Pocket Calculator*,
DOI 10.1007/978-3-319-27104-0_39

| Outcome binary | Predictor binary |
|---|---|
| . | . |
| . | . |
| . | . |
| | |
| | |
| | |

## 3 Primary Scientific Question

Is the Fisher exact test a reliable alternative to the chi-square test and z-test.

## 4 Data Example

The underneath example shows two groups assessed for narcolepsia during the dag.

| | Sleepiness | no sleepiness |
|---|---|---|
| Left treatment (left group) | 5 (a) | 10 (b) |
| Right treatment (right group) | 9 (c) | 6 (d) |

Unlike the chi-square test a z-test, the Fisher test does not make use of chi-square or normal frequency distributions to approximate the level of statistical significance, but, instead, computes exact p-values like 0.05132 (rather than $< 0.05$). The underneath computation is given.

$$\text{p-value} = \text{probability} = \frac{(a+b)!\,((c+d)!\,(a+c)!\,(b+d))!}{(a+b+c+d)!\,a!b!c!d!} = 0.200$$

The chi-square value from the above data is in the previous chapter (Chap. 38), and equals 2.143. A approximated p-value as obtained from the t-table is $>0.10$. A more precise approximation can be obtained from the internet. E.g., the Quick P Value from Chi-Square Score Calculator is helpful. The approximated p-value from the internet = 0.143222. This is much larger than 0.05, but considerably smaller than 0.200. Fisher-exact test may be OK, but it is, obviously, somewhat conservative as compared to the traditional chi-square test. This means that statistical significance tends to be somewhat harder to obtain.

# 5 Conclusion

Fisher-exact test is used as a test for the analysis of cross-tabs, and also as a contrast test to the chi-square test (Chap. 38) and the z-test (Chap. 36). It is, particularly, convenient for small samples, e.g., samples of $n < 100$.

The approximated p-value from the chi-square test and z-test tend to be smaller than those of the Fisher-exact test. Consequently, statistical significance may be somewhat harder to obtain with the Fisher-exact test.

# 6 Note

More background, theoretical and mathematical information of the Fisher-exact test is given in Statistics applied to clinical studies 5th edition, Chap. 3, Springer Heidelberg Germany, 2012, from the same authors.

# Chapter 40
# Confounding

## 1 General Purpose

In the Chaps. 23 and 24 confounding of continuous outcome data has been assessed. Briefly, with confounding the treatment efficacies are better in one subgroup than they are in the other. For binary data ( = proportional data) this means that the proportions of responders in one subgroups are better than they are in the other subgroup. In the current chapter we will assess confounding of binary outcome data instead of continuous outcome data.

## 2 Schematic Overview of Type of Data

| Treatment modality | outcome (binary) |
|---|---|
| (0 and 1) | (1 and 2) |
| . | . |
| . | . |
| . | . |
| . | . |
| . | . |
| . | . |
| . | . |
| . | . |
| | |
| | |
| | |



T.J. Cleophas, A.H. Zwinderman, *Clinical Data Analysis on a Pocket Calculator*,
DOI 10.1007/978-3-319-27104-0_40

## 3 Primary Scientific Question

Do treatment efficacies perform better in one subgroup than in the other.

## 4 Data Example, Demonstrating Confounding

The numbers of responders to two different treatments is assessed in a parallel-group study of 384 patients.

| | responders | non-responders | total | proportion | SE (=√[p(1−p)/n)] |
|---|---|---|---|---|---|
| males: | | | | | |
| treat 1 | 36 | 50 | 86 | 36/86 = 0.42 | √(0.42 × 0.58/86) = 0.05 |
| treat 2 | 14 | 50 | 64 | 14/64 = 0.22 | = 0.05 |
| total | 50 | 100 | 150 | | |
| females: | | | | | |
| treat 1 | 24 | 10 | 34 | 0.71 | =0.08 |
| treat 2 | 120 | 80 | 200 | 0.60 | =0.03 |
| total | 144 | 90 | 234 | | |
| together: | | | | | |
| treat 1 | 60 | 60 | 120 | | |
| treat 2 | 134 | 130 | 264 | | |
| total | 194 | 190 | 384 | | |

p = proportion, n = sample size, SE = standard error

For the males the treatments 1 and 2 perform significantly different, because

t = difference of proportions/its pooled variance =
$(0.42 - 0.22)/(SE_{treat\ 1}{}^2 + SE_{treat\ 2}{}^2) =$
$0.20/0.07 = 2.86.$

With proportional data t-values are more often called z-values.

| df | One-Tail = .4<br>Two-Tail = .8 | .25<br>.5 | .1<br>.2 | .05<br>.1 | .025<br>.05 | .01<br>.02 | .005<br>.01 | .0025<br>.005 | .001<br>.002 | .0005<br>.001 |
|---|---|---|---|---|---|---|---|---|---|---|
| 1 | 0.325 | 1.000 | 3.078 | 6.314 | 12.706 | 31.821 | 63.657 | 127.32 | 318.31 | 636.62 |
| 2 | 0.289 | 0.816 | 1.886 | 2.920 | 4.303 | 6.965 | 9.925 | 14.089 | 22.327 | 31.598 |
| 3 | 0.277 | 0.765 | 1.638 | 2.353 | 3.182 | 4.541 | 5.841 | 7.453 | 10.214 | 12.924 |
| 4 | 0.271 | 0.741 | 1.533 | 2.132 | 2.776 | 3.747 | 4.604 | 5.598 | 7.173 | 8.610 |
| 5 | 0.267 | 0.727 | 1.476 | 2.015 | 2.571 | 3.365 | 4.032 | 4.773 | 5.893 | 6.869 |
| 6 | 0.265 | 0.718 | 1.440 | 1.943 | 2.447 | 3.143 | 3.707 | 4.317 | 5.208 | 5.959 |
| 7 | 0.263 | 0.711 | 1.415 | 1.895 | 2.365 | 2.998 | 3.499 | 4.029 | 4.785 | 5.408 |
| 8 | 0.262 | 0.706 | 1.397 | 1.860 | 2.306 | 2.896 | 3.355 | 3.833 | 4.501 | 5.041 |
| 9 | 0.261 | 0.703 | 1.383 | 1.833 | 2.262 | 2.821 | 3.250 | 3.690 | 4.297 | 4.781 |
| 10 | 0.260 | 0.700 | 1.372 | 1.812 | 2.228 | 2.764 | 3.169 | 3.581 | 4.144 | 4.587 |
| 11 | 0.260 | 0.697 | 1.363 | 1.796 | 2.201 | 2.718 | 3.106 | 3.497 | 4.025 | 4.437 |
| 12 | 0.259 | 0.695 | 1.356 | 1.782 | 2.179 | 2.681 | 3.055 | 3.428 | 3.930 | 4.318 |
| 13 | 0.259 | 0.694 | 1.350 | 1.771 | 2.160 | 2.650 | 3.012 | 3.372 | 3.852 | 4.221 |
| 14 | 0.258 | 0.692 | 1.345 | 1.761 | 2.145 | 2.624 | 2.977 | 3.326 | 3.787 | 4.140 |
| 15 | 0.258 | 0.691 | 1.341 | 1.753 | 2.131 | 2.602 | 2.947 | 3.286 | 3.733 | 4.073 |
| 16 | 0.258 | 0.690 | 1.337 | 1.746 | 2.120 | 2.583 | 2.921 | 3.252 | 3.686 | 4.015 |
| 17 | 0.257 | 0.689 | 1.333 | 1.740 | 2.110 | 2.567 | 2.898 | 3.222 | 3.646 | 3.965 |
| 18 | 0.257 | 0.688 | 1.330 | 1.734 | 2.101 | 2.552 | 2.878 | 3.197 | 3.610 | 3.922 |
| 19 | 0.257 | 0.688 | 1.328 | 1.729 | 2.093 | 2.539 | 2.861 | 3.174 | 3.579 | 3.883 |
| 20 | 0.257 | 0.687 | 1.325 | 1.725 | 2.086 | 2.528 | 2.845 | 3.153 | 3.552 | 3.850 |
| 21 | 0.257 | 0.686 | 1.323 | 1.721 | 2.080 | 2.518 | 2.831 | 3.135 | 3.527 | 3.819 |
| 22 | 0.256 | 0.686 | 1.321 | 1.717 | 2.074 | 2.508 | 2.819 | 3.119 | 3.505 | 3.792 |
| 23 | 0.256 | 0.685 | 1.319 | 1.714 | 2.069 | 2.500 | 2.807 | 3.104 | 3.485 | 3.767 |
| 24 | 0.256 | 0.685 | 1.318 | 1.711 | 2.064 | 2.492 | 2.797 | 3.091 | 3.467 | 3.745 |
| 25 | 0.256 | 0.684 | 1.316 | 1.708 | 2.060 | 2.485 | 2.787 | 3.078 | 3.450 | 3.725 |
| 26 | 0.256 | 0.684 | 1.315 | 1.706 | 2.056 | 2.479 | 2.779 | 3.067 | 3.435 | 3.707 |
| 27 | 0.256 | 0.684 | 1.314 | 1.703 | 2.052 | 2.473 | 2.771 | 3.057 | 3.421 | 3.690 |
| 28 | 0.256 | 0.683 | 1.313 | 1.701 | 2.048 | 2.467 | 2.763 | 3.047 | 3.408 | 3.674 |
| 29 | 0.256 | 0.683 | 1.311 | 1.699 | 2.045 | 2.462 | 2.756 | 3.038 | 3.396 | 3.659 |
| 30 | 0.256 | 0.683 | 1.310 | 1.697 | 2.042 | 2.457 | 2.750 | 3.030 | 3.385 | 3.646 |
| 40 | 0.255 | 0.681 | 1.303 | 1.684 | 2.021 | 2.423 | 2.704 | 2.971 | 3.307 | 3.551 |
| 60 | 0.254 | 0.679 | 1.296 | 1.671 | 2.000 | 2.390 | 2.660 | 2.915 | 3.232 | 3.460 |
| 120 | 0.254 | 0.677 | 1.289 | 1.658 | 1.980 | 2.358 | 2.617 | 2.860 | 3.160 | 3.373 |
| ∞ | 0.253 | 0.674 | 1.282 | 1.645 | 1.960 | 2.326 | 2.576 | 2.807 | 3.090 | 3.291 |

The t-table has a left-end column giving degrees of freedom (≈ sample sizes), and two top rows with p-values (areas under the curve = p – values), one-tail meaning that only one end of the curve, two-tail meaning that both ends are assessed simultaneously. The t-table is, furthermore, full of t-values, that, with ∞ degrees of freedom, are equal to z-values. The z-values and t-values are to be understood as mean results of studies, but not expressed in mmol/l, kilograms, or proportions of responders, but in so-called SEM-units (Standard error of the mean units), that are obtained by dividing your mean result by its own standard error. For continuous outcome data, with many degrees of freedom (large samples) the curve will be a little bit narrower, and more in agreement with nature. For binary outcome data, nature has determined that the curves will always be as narrow as can be, according to the row at the bottom.

A z-value of 2.86 is larger than 2.807, and, therefore, statistically significant from zero with a two-tail p-value <0.05.

For the females the treatments 1 and 2 do not perform significantly differently, because

$$
\begin{aligned}
z = \; & \text{difference of proportions/its pooled variance} = \\
& (0.71 - 0.60)/(SE_{treat\ 1}{}^2 + SE_{treat\ 2}{}^2) = \\
& 0.11/0.09 = 1.22.
\end{aligned}
$$

This difference is smaller than 1.960, and therefore, not statistically significantly better than a difference of 0.

For the combined data, the treatments 1 and 2 do not perform significantly differently from zero, because

$$
\begin{aligned}
z = \; & \text{difference of proportions/its pooled variance} = \\
& (0.50 - 0.51)/(SE_{treat\ 1}{}^2 + SE_{treat\ 2}{}^2) = \\
& 0.01/.... = \text{very small.}
\end{aligned}
$$

This difference is again not statistically significant from a difference of zero, and, thus, the treatments do not perform significantly differently from one another.

## 5 Testing Confounding with a Z-Test

A weighted mean proportion is calculated, and tested (variance $= SE^2$). The underneath differences indicate, respectively, the differences in the males and the females.

$$
\frac{\text{difference/variance} \quad + \text{difference/variance}}{1/\text{variance} \quad + 1/\text{variance}} = \frac{0.20/0.05^2 + 0.11/0.09^2}{1/0.05^2 + 1/0.09^2} = \frac{80 + 13.6}{4000 + 123.5} = \frac{93.6}{523.5} = 0.18
$$

This weighted mean proportion of 0.18 is much closer to 0.20 than to 0.11, and this is due to the much larger sample size of females than that of the males. We will now test the weighted mean proportion against its SE (standard error).

$$
\begin{aligned}
SE &= \sqrt{}\,[1/(1/SE_1{}^2 + 1/SE_2{}^2)] = 0.044 \\
z &= \text{weighted mean/its SE} = 0.18/0.044 = 4.9
\end{aligned}
$$

This z-value is larger than 3.291, and, therefore, produces a two-tail p-value of <0.01, and, so, after adjustment for confounding between males and females, the treatments 1 and 2 perform very significantly different from one another.

## 6 Testing Confounding with a Mantel-Haenszl Test

The Mantel-Haenszl chi-square test is equivalent to the z-test.

males

| observed | expected | variance ( n = 150) |
|---|---|---|
| 36 | $(86 \times 50)/150 = 28.7$ | $86 \times 64 \times 50 \times 100/n^2(n-1) = 8.21$ |

females

| observed | expected | variance ( n = 234) |
|---|---|---|
| 24 | $(34 \times 144)/234 = 20.9$ | $34 \times 200 \times 144 \times 90/n^2(n-1) = 6.91.$ |

Chi-square $= (28.7 - 20.9)^2/(\text{var } 1 + \text{var } 2) = 7.8^2/15.12 = 4.02$

With 1 degree of freedom the p-value should be $< 0.01$ (var = variance).

The underneath chi-square table has an upper row with areas under the curve, a left-end column with degrees of freedom, and a whole lot of chi-square values.

The chi-square value of 4.02 with one degree of freedom is larger than 3.841, and, thus the two-tail p-value is $<0.05$.

Chi-squared distribution

| | Two-tailed *P*-value | | | |
|---|---|---|---|---|
| *df* | 0.10 | 0.05 | 0.01 | 0.001 |
| 1 | 2.706 | 3.841 | 6.635 | 10.827 |
| 2 | 4.605 | 5.991 | 9.210 | 13.815 |
| 3 | 6.251 | 7.851 | 11.345 | 16.266 |
| 4 | 7.779 | 9.488 | 13.277 | 18.466 |
| 5 | 9.236 | 11.070 | 15.086 | 20.515 |
| 6 | 10.645 | 12.592 | 16.812 | 22.457 |
| 7 | 12.017 | 14.067 | 18.475 | 24.321 |
| 8 | 13.362 | 15.507 | 20.090 | 26.124 |
| 9 | 14.684 | 16.919 | 21.666 | 27.877 |
| 10 | 15.987 | 18.307 | 23.209 | 29.588 |
| 11 | 17.275 | 19.675 | 24.725 | 31.264 |
| 12 | 18.549 | 21.026 | 26.217 | 32.909 |
| 13 | 19.812 | 22.362 | 27.688 | 34.527 |
| 14 | 21.064 | 23.685 | 29.141 | 36.124 |
| 15 | 22.307 | 24.996 | 30.578 | 37.698 |
| 16 | 23.542 | 26.296 | 32.000 | 39.252 |
| 17 | 24.769 | 27.587 | 33.409 | 40.791 |
| 18 | 25.989 | 28.869 | 34.805 | 42.312 |
| 19 | 27.204 | 30.144 | 36.191 | 43.819 |
| 20 | 28.412 | 31.410 | 37.566 | 45.314 |
| 21 | 29.615 | 32.671 | 38.932 | 46.796 |
| 22 | 30.813 | 33.924 | 40.289 | 48.268 |
| 23 | 32.007 | 35.172 | 41.638 | 49.728 |

(continued)

| | Two-tailed *P*-value | | | |
|---|---|---|---|---|
| *df* | 0.10 | 0.05 | 0.01 | 0.001 |
| 24 | 33.196 | 36.415 | 42.980 | 51.179 |
| 25 | 34.382 | 37.652 | 44.314 | 52.619 |
| 26 | 35.536 | 38.885 | 45.642 | 54.051 |
| 27 | 36.741 | 40.113 | 46.963 | 55.475 |
| 28 | 37.916 | 41.337 | 48.278 | 56.892 |
| 29 | 39.087 | 42.557 | 49.588 | 58.301 |
| 30 | 40.256 | 43.773 | 50.892 | 59.702 |
| 40 | 51.805 | 55.758 | 63.691 | 73.403 |
| 50 | 63.167 | 67.505 | 76.154 | 86.660 |
| 60 | 74.397 | 79.082 | 88.379 | 99.608 |
| 70 | 85.527 | 90.531 | 100.43 | 112.32 |
| 80 | 96.578 | 101.88 | 112.33 | 124.84 |
| 90 | 107.57 | 113.15 | 124.12 | 137.21 |
| 100 | 118.50 | 124.34 | 135.81 | 149.45 |

## 7 Conclusion

Both z-test and Mantel-Haenszl chi-square test results show that after adjustment for confounding the non-significant difference between the effects of the treatments 1 and 2 turn into significant effects with p-values of <0.01 and <0.05.

## 8 Note

More background, theoretical and mathematical information of confounding assessments can be found in Statistics applied top clinical studies 5th edition, Chap. 28, Springer Heidelberg Germany, from the same authors.

# Chapter 41
# Interaction

## 1 General Purpose

In the Chap. 25 interaction of continuous outcome data has been assessed. The efficacy of one treatment is better in one of the subgroups, and the efficacy of the other treatment is better in the other subgroup. With interaction an overall data analysis is pretty meaningless, and separate analyses of the subgroups must be presented. Of course, with multiple treatments and subgroups interaction is possible as well, but the analysis is much more complex. In the current chapter we will demonstrate how to assess interaction of binary outcome data instead of continuous outcome data.

## 2 Schematic Overview of Type of Data

| Treatment modality | outcome (binary) |
|---|---|
| (0 and 1) | (1 and 2) |
| . | . |
| . | . |
| . | . |
| . | . |
| . | . |
| . | . |
| . | . |
| . | . |
| | |
| | |
| | |



T.J. Cleophas, A.H. Zwinderman, *Clinical Data Analysis on a Pocket Calculator*,
DOI 10.1007/978-3-319-27104-0_41

## 3 Primary Scientific Question

Does one treatment perform better in one subgroup and does the other treatment so in the other subgroup.

## 4 Interaction, Example 1

We will use the example of the Chap. 40, Sect. 40.4, once more to assess the presence of interaction between males and females on the outcome.

| | males | females |
|---|---|---|
| proportion treat 1 | 0.42 | 0.71 |
| SE | 0.05 | 0.08 |
| proportion treat 2 | 0.22 | 0.60 |
| SE | 0.05 | 0.03 |
| Differences | 0.20 | 0.11 |
| pooled SE | 0.07 | 0.09 |

Differences between males and females 0.20−0.11 = 0.09 with a pooled SE of $\sqrt{[(0.07^2) + (0.09^2)]} = 0.11$. Z-value = 0.09/0.11 = 0.82. This z-value not statistically significantly different from zero (see the underneath t-table). It means that, although confounding of genders has been demonstrated in these data (Chap. 40), no interaction between the genders is in the data.

| df | One-Tail = .4<br>Two-Tail = .8 | .25<br>.5 | .1<br>.2 | .05<br>.1 | .025<br>.05 | .01<br>.02 | .005<br>.01 | .0025<br>.005 | .001<br>.002 | .0005<br>.001 |
|---|---|---|---|---|---|---|---|---|---|---|
| 1 | 0.325 | 1.000 | 3.078 | 6.314 | 12.706 | 31.821 | 63.657 | 127.32 | 318.31 | 636.62 |
| 2 | 0.289 | 0.816 | 1.886 | 2.920 | 4.303 | 6.965 | 9.925 | 14.089 | 22.327 | 31.598 |
| 3 | 0.277 | 0.765 | 1.638 | 2.353 | 3.182 | 4.541 | 5.841 | 7.453 | 10.214 | 12.924 |
| 4 | 0.271 | 0.741 | 1.533 | 2.132 | 2.776 | 3.747 | 4.604 | 5.598 | 7.173 | 8.610 |
| 5 | 0.267 | 0.727 | 1.476 | 2.015 | 2.571 | 3.365 | 4.032 | 4.773 | 5.893 | 6.869 |
| 6 | 0.265 | 0.718 | 1.440 | 1.943 | 2.447 | 3.143 | 3.707 | 4.317 | 5.208 | 5.959 |
| 7 | 0.263 | 0.711 | 1.415 | 1.895 | 2.365 | 2.998 | 3.499 | 4.029 | 4.785 | 5.408 |
| 8 | 0.262 | 0.706 | 1.397 | 1.860 | 2.306 | 2.896 | 3.355 | 3.833 | 4.501 | 5.041 |
| 9 | 0.261 | 0.703 | 1.383 | 1.833 | 2.262 | 2.821 | 3.250 | 3.690 | 4.297 | 4.781 |
| 10 | 0.260 | 0.700 | 1.372 | 1.812 | 2.228 | 2.764 | 3.169 | 3.581 | 4.144 | 4.587 |
| 11 | 0.260 | 0.697 | 1.363 | 1.796 | 2.201 | 2.718 | 3.106 | 3.497 | 4.025 | 4.437 |
| 12 | 0.259 | 0.695 | 1.356 | 1.782 | 2.179 | 2.681 | 3.055 | 3.428 | 3.930 | 4.318 |
| 13 | 0.259 | 0.694 | 1.350 | 1.771 | 2.160 | 2.650 | 3.012 | 3.372 | 3.852 | 4.221 |
| 14 | 0.258 | 0.692 | 1.345 | 1.761 | 2.145 | 2.624 | 2.977 | 3.326 | 3.787 | 4.140 |
| 15 | 0.258 | 0.691 | 1.341 | 1.753 | 2.131 | 2.602 | 2.947 | 3.286 | 3.733 | 4.073 |
| 16 | 0.258 | 0.690 | 1.337 | 1.746 | 2.120 | 2.583 | 2.921 | 3.252 | 3.686 | 4.015 |
| 17 | 0.257 | 0.689 | 1.333 | 1.740 | 2.110 | 2.567 | 2.898 | 3.222 | 3.646 | 3.965 |
| 18 | 0.257 | 0.688 | 1.330 | 1.734 | 2.101 | 2.552 | 2.878 | 3.197 | 3.610 | 3.922 |
| 19 | 0.257 | 0.688 | 1.328 | 1.729 | 2.093 | 2.539 | 2.861 | 3.174 | 3.579 | 3.883 |
| 20 | 0.257 | 0.687 | 1.325 | 1.725 | 2.086 | 2.528 | 2.845 | 3.153 | 3.552 | 3.850 |
| 21 | 0.257 | 0.686 | 1.323 | 1.721 | 2.080 | 2.518 | 2.831 | 3.135 | 3.527 | 3.819 |
| 22 | 0.256 | 0.686 | 1.321 | 1.717 | 2.074 | 2.508 | 2.819 | 3.119 | 3.505 | 3.792 |
| 23 | 0.256 | 0.685 | 1.319 | 1.714 | 2.069 | 2.500 | 2.807 | 3.104 | 3.485 | 3.767 |
| 24 | 0.256 | 0.685 | 1.318 | 1.711 | 2.064 | 2.492 | 2.797 | 3.091 | 3.467 | 3.745 |
| 25 | 0.256 | 0.684 | 1.316 | 1.708 | 2.060 | 2.485 | 2.787 | 3.078 | 3.450 | 3.725 |
| 26 | 0.256 | 0.684 | 1.315 | 1.706 | 2.056 | 2.479 | 2.779 | 3.067 | 3.435 | 3.707 |
| 27 | 0.256 | 0.684 | 1.314 | 1.703 | 2.052 | 2.473 | 2.771 | 3.057 | 3.421 | 3.690 |
| 28 | 0.256 | 0.683 | 1.313 | 1.701 | 2.048 | 2.467 | 2.763 | 3.047 | 3.408 | 3.674 |
| 29 | 0.256 | 0.683 | 1.311 | 1.699 | 2.045 | 2.462 | 2.756 | 3.038 | 3.396 | 3.659 |
| 30 | 0.256 | 0.683 | 1.310 | 1.697 | 2.042 | 2.457 | 2.750 | 3.030 | 3.385 | 3.646 |
| 40 | 0.255 | 0.681 | 1.303 | 1.684 | 2.021 | 2.423 | 2.704 | 2.971 | 3.307 | 3.551 |
| 60 | 0.254 | 0.679 | 1.296 | 1.671 | 2.000 | 2.390 | 2.660 | 2.915 | 3.232 | 3.460 |
| 120 | 0.254 | 0.677 | 1.289 | 1.658 | 1.980 | 2.358 | 2.617 | 2.860 | 3.160 | 3.373 |
| ∞ | 0.253 | 0.674 | 1.282 | 1.645 | 1.960 | 2.326 | 2.576 | 2.807 | 3.090 | 3.291 |

The t-table has a left-end column giving degrees of freedom (≈ sample sizes), and two top rows with p-values (areas under the curve = p – values), one-tail meaning that only one end of the curve, two-tail meaning that both ends are assessed simultaneously. The t-table is, furthermore, full of t-values, that, with ∞ degrees of freedom, are equal to z-values (Chap. 36). The t-values are to be understood as mean results of studies, but not expressed in mmol/l, kilograms, but in so-called SEM-units (Standard error of the mean units), that are obtained by dividing your mean result by its own standard error. With many degrees of freedom (large samples) the curve will be a little bit narrower, and more in agreement with nature.

A z-value of 0.82 is a lot smaller than 1.960, and the corresponding two-tail p-value is, thus, > 0.05, and not significant.

## 5 Interaction, Example 2

The hypothesized data from the above example have been slightly changed.

| | responders | non-responders | total | proportion | $SE = \sqrt{[p(1-p)/n)]}$ |
|---|---|---|---|---|---|
| males: | | | | | |
| treat 1 | 36 | 50 | 86 | 36/86 = 0.42 | $\sqrt{(0.42 \times 0.58/86)}$ =0.05 |
| treat 2 | 14 | 50 | 64 | 14/64 = 0.22 | =0.05 |
| total | 50 | 100 | 150 | | |
| females: | | | | | |
| treat 1 | 120 | 80 | 200 | 0.60 | =0.03 |
| treat 2 | 24 | 10 | 34 | 0.71 | =0.08 |
| total | 144 | 90 | 234 | | |

p = proportion, n = sample size, SE = standard error

| | males | females |
|---|---|---|
| proportion treat 1 | 0.42 | 0.60 |
| SE | 0.05 | 0.03 |
| proportion treat 2 | 0.22 | 0.71 |
| SE | 0.05 | 0.08 |
| Differences | 0.20 | −0.11 |
| pooled SE | 0.07 | 0.09 |

Difference in proportion between males and females 0.20 + 0.11 = 0.31 with a pooled SE of $\sqrt{[(0.07^2) + (0.09^2)]} = 0.11$. The z-value = 0.31/0.11 = 2.82. This z-value statistically larger than 1.960, the p-value is significantly different from zero (see above t-table). It means that a significant interaction between genders is in these data. In the males the treatment 1 performs better, in the females the treatment 2.

# 6 Conclusion

A two-sample t-test (or rather z-test) (see also Chap. 36) can be used to test whether the differences in treatment efficacies of two subgroups are significantly different from one another. The above example shows that a significant interaction between genders can be demonstrated in such data. In the males the treatment 1 performed better, in the females the treatment 2 did so.

# 7 Note

More background, theoretical and mathematical information of interaction assessments can be found in Statistics applied top clinical studies 5th edition, Chap. 30, Springer Heidelberg Germany, from the same authors.

# Chapter 42
# Chi-Square Tests for Large Cross-Tabs

## 1 General Purpose

Chi-square tests are adequate for testing $2 \times 2$ interaction cross-tabs of two treatment modalities and two numbers of responders to treatment (Chap. 38). These tests can, however, equally well, be applied for testing larger tables.

## 2 Schematic Overview of Type of Data File

| Predictor (1–3...) | outcome (binary) |
|---|---|
| . | . |
| . | . |
| . | . |
| . | . |
| . | . |
| . | . |
| . | . |
| | |
| | |
| | |
| | |



T.J. Cleophas, A.H. Zwinderman, *Clinical Data Analysis on a Pocket Calculator*,
DOI 10.1007/978-3-319-27104-0_42

## 3 Primary Scientific Question

In a $3 \times 2$ interaction cross-tab, is there a significant difference between three treatment modalities in numbers of responders to treatment?

## 4 Data Example One: $3 \times 2$ Cross-Tab

In three different treatment groups of hypertensive patients the numbers of responders (normotensive after treatment) is assessed.

| | Responders yes | no | total |
|---|---|---|---|
| Treatment 1 | 60 | 40 | 100 |
| Treatment 2 | 100 | 120 | 220 |
| Treatment 3 | 80 | 60 | 140 |
| total | 240 | 220 | 460 |

The best estimate of expectation is calculated from the above data under the null-hypothesis, that the results are not significantly different between the groups. E.g., estimate expected number of responders in treatment group 1 by dividing all responders (240) by all observations (460), and multiply by observations in treatment group 1 (100). Then, this estimate (the expected numbers E) is compared with the actually observed numbers of responders (observed numbers O). The procedure is illustrated below for responders and, also, non-responders in the treatment 1 group.

| | Expected (E) Responders yes | no | O−E | $(O-E)^2/E$ |
|---|---|---|---|---|
| Treatment 1 | $(240/460) \times 100$ | $(220/460) \times 100$ | ... | ... |
| Treatment 2 | ... | ... | ... | ... |
| Treatment 3 | ... | ... | ... | ... |

The add-up sum of the above three $(O-E)^2/E$ terms equal the chi-square value. The p-value can be read from the chi-square table for $3-1=2$ degrees of freedom.

The above procedure is laborious, and a *fast* method producing the same result is given underneath.

$240^2/460 = \quad 125.22$

Subtract the above value from the add-up sum of the underneath values.

$60^2/100 = \quad 36.00$
$100^2/220 = \quad 45.45$
$80^2/140 = \quad 45.71$

The result equals 1.9…

The chi-square value =

$$1.9\ldots/[(240/460) \times (220/460)] = 7.19\ldots$$

The underneath chi-square table has an upper row with areas under the curve, a left-end column with degrees of freedom, and a whole lot of chi-square values. The chi-square table shows that with 3 – 1 = 2 degrees of freedom (second row) 7.19… is between 5.991 and 9.210, and, thus, that the corresponding p-value (in the top-row) is between 0.05 and 0.01.

Chi-squared distribution

| | Two-tailed *P*-value | | | |
|---|---|---|---|---|
| *df* | 0.10 | 0.05 | 0.01 | 0.001 |
| 1 | 2.706 | 3.841 | 6.635 | 10.827 |
| 2 | 4.605 | 5.991 | 9.210 | 13.815 |
| 3 | 6.251 | 7.851 | 11.345 | 16.266 |
| 4 | 7.779 | 9.488 | 13.277 | 18.466 |
| 5 | 9.236 | 11.070 | 15.086 | 20.515 |
| 6 | 10.645 | 12.592 | 16.812 | 22.457 |
| 7 | 12.017 | 14.067 | 18.475 | 24.321 |
| 8 | 13.362 | 15.507 | 20.090 | 26.124 |
| 9 | 14.684 | 16.919 | 21.666 | 27.877 |
| 10 | 15.987 | 18.307 | 23.209 | 29.588 |
| 11 | 17.275 | 19.675 | 24.725 | 31.264 |
| 12 | 18.549 | 21.026 | 26.217 | 32.909 |
| 13 | 19.812 | 22.362 | 27.688 | 34.527 |
| 14 | 21.064 | 23.685 | 29.141 | 36.124 |
| 15 | 22.307 | 24.996 | 30.578 | 37.698 |
| 16 | 23.542 | 26.296 | 32.000 | 39.252 |
| 17 | 24.769 | 27.587 | 33.409 | 40.791 |
| 18 | 25.989 | 28.869 | 34.805 | 42.312 |
| 19 | 27.204 | 30.144 | 36.191 | 43.819 |
| 20 | 28.412 | 31.410 | 37.566 | 45.314 |
| 21 | 29.615 | 32.671 | 38.932 | 46.796 |
| 22 | 30.813 | 33.924 | 40.289 | 48.268 |
| 23 | 32.007 | 35.172 | 41.638 | 49.728 |
| 24 | 33.196 | 36.415 | 42.980 | 51.179 |
| 25 | 34.382 | 37.652 | 44.314 | 52.619 |
| 26 | 35.536 | 38.885 | 45.642 | 54.051 |
| 27 | 36.741 | 40.113 | 46.963 | 55.475 |
| 28 | 37.916 | 41.337 | 48.278 | 56.892 |
| 29 | 39.087 | 42.557 | 49.588 | 58.301 |
| 30 | 40.256 | 43.773 | 50.892 | 59.702 |

(continued)

| | Two-tailed *P*-value | | | |
|---|---|---|---|---|
| *df* | 0.10 | 0.05 | 0.01 | 0.001 |
| 40 | 51.805 | 55.758 | 63.691 | 73.403 |
| 50 | 63.167 | 67.505 | 76.154 | 86.660 |
| 60 | 74.397 | 79.082 | 88.379 | 99.608 |
| 70 | 85.527 | 90.531 | 100.43 | 112.32 |
| 80 | 96.578 | 101.88 | 112.33 | 124.84 |
| 90 | 107.57 | 113.15 | 124.12 | 137.21 |
| 100 | 118.50 | 124.34 | 135.81 | 149.45 |

Obviously, the three treatment modalities are significantly different from one another. In order to find out, whether the significant effect is between the treatments 1 and 2, 2 and 3, or 1 and 3, subsequent post hoc tests must be performed using $2 \times 2$ chi-square tests (Chap. 38). Bonferroni adjustments are to be recommended (Chap. 18).

## 5 Data Example Two: Theoretical Distribution

A random sample of 200 subjects was assessed for sleepiness during the day. We have been given demographic information (theoretical distribution) about the prevalence of sleepiness in the population from which the sample was taken. We wish to know whether the sample's distribution is not different from that of the population.

| | Expected (theoretically) | Observed | O – E |
|---|---|---|---|
| 1.sleepy | 0.24 | 64 | $64 - (0.24 \times 200) = 16$ |
| 2.sleepy rarely | 0.60 | 124 | $124 - (0.6 \times 200) = 4$ |
| 3.sleepy never | 0.16 | 12 | $12 - (0.16 \times 200) = -20$ |

| | $(O-E)^2/E$ | | |
|---|---|---|---|
| 1.sleepy | $16^2/(0.24 \times 200)$ | = | 5.33 |
| 2.sleepy rarely | $4^2/(0.60 \times 200)$ | = | 0.05 |
| 3.sleepy never | $(-20^2)/(0.16 \times 200)$ | = | 12.50 |
| total | | | 17.88 |

According to the chi-square value of 17.88 with $3 - 1 = 2$ degrees of freedom, which is larger than 13.815, the p-value should $<0.001$ (see chi-square table in the above Sect. 42.4). Many more sleepy people were in the observed population than expected from the theoretical distribution. And, so, they must have had some reason for being so sleepy, that should be searched for properly.

## 6 Conclusion

Chi-square tests are adequate for testing 2 × 2 interaction cross-tabs of two treatment modalities and two numbers of responders to treatment. They can, however, equally well be applied for testing larger tables. In clinical research 2 × 2 tables are far more commonly used than large tables, because clinicians are mostly interested to find a single best treatment rather than a significant difference somewhere in multiple treatments. After an overall assessment of a larger table, multiple Bonferroni -adjusted subsequent 2 × 2 tests are required to find out this single best treatment. You may consider to skip the overall assessment and start with 2 × 2 tests from the very beginning.

## 7 Note

More background, theoretical and mathematical information of large chi-square tables are given in Statistics applied to clinical studies 5th edition, Chap. 3, Springer Heidelberg Germany, from the same authors.

# Chapter 43
# Logarithmic Transformations, a Great Help to Statistical Analyses

## 1 General Purpose

Non-linear relationships in clinical research are often linear after logarithmic transformations. Also, logarithmic transformation normalizes skewed frequency distributions and is used for the analysis of likelihood ratios. Basic knowledge of logarithms is, therefore, convenient for a better understanding of many statistical methods. Almost always natural logarithm (ln), otherwise called Naperian logarithm, is used, i.e., logarithm to the base e. Log is logarithm to the base 10, ln is logarithm to the base e (2.718281828). This chapter is for showing that knowledge of logarithms is helpful for understanding many statistical methods.

## 2 Schematic Overview of Type of Data File

| Outcome binary | predictor binary |
|---|---|
| . | . |
| . | . |
| . | . |
| . | . |
| . | . |
| . | . |
| . | . |
| . | . |
| . | . |
| | |
| | |
| | |



T.J. Cleophas, A.H. Zwinderman, *Clinical Data Analysis on a Pocket Calculator*,
DOI 10.1007/978-3-319-27104-0_43

## 3 Primary Scientific Question

Is logarithmic transformation relevant for statistical modeling.

## 4 Theory and Basic Steps

log 10 = 10 log 10 = 1
log 100 = 10 log 100 = 2
log 1 = 10 log 1 = 0
antilog 1 = 10
antilog 2 = 100
antilog 0 = 1

Casio fx-825 scientific, Scientific Calculator, Texas TI-30XA, Sigma, Commodore
Press: 100….log….2
Press: 2….2ndf….log…100

Electronic Calculator, Kenko KK-82MS-5
Press: 100…. = ….log…. = ….2
Press: 2…. = ….shift…log….100

ln e = e log e = 1
ln $e^2$ = e log $e^2$ = 2
ln 1 = e log 1 = 0
antiln 1 = 2.718…
antiln 2 = 7.389…
antiln 0 = 1

Casio fx-825 scientific, Scientific Calculator, Texas TI-30XA, Sigma
Press: 7.389….ln….2
Press: 2….2ndf….ln…7389

Electronic Calculator, Kenko KK-82MS-5
Press: 7.389…. = ….ln…. = ….2
Press: 2…. = ….shift…ln….7.389

## 5 Example, Markov Model

In patients with diabetes mellitus (* = sign of multiplication):

After 1 year 10 % has beta-cell failure, and 90 % has not.
2 90 * 90 = 81 % has not.
3 90 * 90 * 90 = 73 % has not.

When will 50 % have beta-cell failure?
$0.9^x = 0.5$
$x \log 0.9 = \log 0.5$
$x = \log 0.5/\log 0.9 = 6.5788$ years.

## 6 Example, Odds Ratios

| | events | no events | |
|---|---|---|---|
| | numbers of patients | | |
| group 1 | 15(a) | 20(b) | 35(a + b) |
| group 2 | 15(c) | 5(d) | 20(c + d) |
| | 30(a + c) | 25(b + d) | 55(a + b + c + d) |

The odds of an event = the number of patients in a group with an event divided by the number without. In group 1 the odds of an event equals = a/b.

The odds ratio (OR) of group 1 compared to group 2
= (a/b)/(c/d)
= (15/20)/(15/5)
= 0.25

lnOR = ln 0.25 = −1.386 (ln = natural logarithm)

The standard error (SE) of the above term
= √ (1/a + 1/b + 1/c + 1/d)
= √ (1/15 + 1/20 + 1/15 + 1/5)
= √ 0.38333
= 0.619

The odds ratio can be tested using the z-test.

The test-statistic = z-value
= (ln odds ratio)/(SE ln odds ratio)
= −1.386/0.619
= −2.239

The underneath t-table, bottom row, shows, that, if this value is smaller than −1.96 or larger than +1.96, then the odds ratio is significantly different from 1 with a two-tail p-value < 0.05. There is, thus, a significant difference in numbers of events between the two groups.

| *df* | One-Tail = .4<br>Two-Tail = .8 | .25<br>.5 | .1<br>.2 | .05<br>.1 | .025<br>.05 | .01<br>.02 | .005<br>.01 | .0025<br>.005 | .001<br>.002 | .0005<br>.001 |
|---|---|---|---|---|---|---|---|---|---|---|
| 1 | 0.325 | 1.000 | 3.078 | 6.314 | 12.706 | 31.821 | 63.657 | 127.32 | 318.31 | 636.62 |
| 2 | 0.289 | 0.816 | 1.886 | 2.920 | 4.303 | 6.965 | 9.925 | 14.089 | 22.327 | 31.598 |
| 3 | 0.277 | 0.765 | 1.638 | 2.353 | 3.182 | 4.541 | 5.841 | 7.453 | 10.214 | 12.924 |
| 4 | 0.271 | 0.741 | 1.533 | 2.132 | 2.776 | 3.747 | 4.604 | 5.598 | 7.173 | 8.610 |
| 5 | 0.267 | 0.727 | 1.476 | 2.015 | 2.571 | 3.365 | 4.032 | 4.773 | 5.893 | 6.869 |
| 6 | 0.265 | 0.718 | 1.440 | 1.943 | 2.447 | 3.143 | 3.707 | 4.317 | 5.208 | 5.959 |
| 7 | 0.263 | 0.711 | 1.415 | 1.895 | 2.365 | 2.998 | 3.499 | 4.029 | 4.785 | 5.408 |
| 8 | 0.262 | 0.706 | 1.397 | 1.860 | 2.306 | 2.896 | 3.355 | 3.833 | 4.501 | 5.041 |
| 9 | 0.261 | 0.703 | 1.383 | 1.833 | 2.262 | 2.821 | 3.250 | 3.690 | 4.297 | 4.781 |
| 10 | 0.260 | 0.700 | 1.372 | 1.812 | 2.228 | 2.764 | 3.169 | 3.581 | 4.144 | 4.587 |
| 11 | 0.260 | 0.697 | 1.363 | 1.796 | 2.201 | 2.718 | 3.106 | 3.497 | 4.025 | 4.437 |
| 12 | 0.259 | 0.695 | 1.356 | 1.782 | 2.179 | 2.681 | 3.055 | 3.428 | 3.930 | 4.318 |
| 13 | 0.259 | 0.694 | 1.350 | 1.771 | 2.160 | 2.650 | 3.012 | 3.372 | 3.852 | 4.221 |
| 14 | 0.258 | 0.692 | 1.345 | 1.761 | 2.145 | 2.624 | 2.977 | 3.326 | 3.787 | 4.140 |
| 15 | 0.258 | 0.691 | 1.341 | 1.753 | 2.131 | 2.602 | 2.947 | 3.286 | 3.733 | 4.073 |
| 16 | 0.258 | 0.690 | 1.337 | 1.746 | 2.120 | 2.583 | 2.921 | 3.252 | 3.686 | 4.015 |
| 17 | 0.257 | 0.689 | 1.333 | 1.740 | 2.110 | 2.567 | 2.898 | 3.222 | 3.646 | 3.965 |
| 18 | 0.257 | 0.688 | 1.330 | 1.734 | 2.101 | 2.552 | 2.878 | 3.197 | 3.610 | 3.922 |
| 19 | 0.257 | 0.688 | 1.328 | 1.729 | 2.093 | 2.539 | 2.861 | 3.174 | 3.579 | 3.883 |
| 20 | 0.257 | 0.687 | 1.325 | 1.725 | 2.086 | 2.528 | 2.845 | 3.153 | 3.552 | 3.850 |
| 21 | 0.257 | 0.686 | 1.323 | 1.721 | 2.080 | 2.518 | 2.831 | 3.135 | 3.527 | 3.819 |
| 22 | 0.256 | 0.686 | 1.321 | 1.717 | 2.074 | 2.508 | 2.819 | 3.119 | 3.505 | 3.792 |
| 23 | 0.256 | 0.685 | 1.319 | 1.714 | 2.069 | 2.500 | 2.807 | 3.104 | 3.485 | 3.767 |
| 24 | 0.256 | 0.685 | 1.318 | 1.711 | 2.064 | 2.492 | 2.797 | 3.091 | 3.467 | 3.745 |
| 25 | 0.256 | 0.684 | 1.316 | 1.708 | 2.060 | 2.485 | 2.787 | 3.078 | 3.450 | 3.725 |
| 26 | 0.256 | 0.684 | 1.315 | 1.706 | 2.056 | 2.479 | 2.779 | 3.067 | 3.435 | 3.707 |
| 27 | 0.256 | 0.684 | 1.314 | 1.703 | 2.052 | 2.473 | 2.771 | 3.057 | 3.421 | 3.690 |
| 28 | 0.256 | 0.683 | 1.313 | 1.701 | 2.048 | 2.467 | 2.763 | 3.047 | 3.408 | 3.674 |
| 29 | 0.256 | 0.683 | 1.311 | 1.699 | 2.045 | 2.462 | 2.756 | 3.038 | 3.396 | 3.659 |
| 30 | 0.256 | 0.683 | 1.310 | 1.697 | 2.042 | 2.457 | 2.750 | 3.030 | 3.385 | 3.646 |
| 40 | 0.255 | 0.681 | 1.303 | 1.684 | 2.021 | 2.423 | 2.704 | 2.971 | 3.307 | 3.551 |
| 60 | 0.254 | 0.679 | 1.296 | 1.671 | 2.000 | 2.390 | 2.660 | 2.915 | 3.232 | 3.460 |
| 120 | 0.254 | 0.677 | 1.289 | 1.658 | 1.980 | 2.358 | 2.617 | 2.860 | 3.160 | 3.373 |
| ∞ | 0.253 | 0.674 | 1.282 | 1.645 | 1.960 | 2.326 | 2.576 | 2.807 | 3.090 | 3.291 |

The left-end column of the above t-table gives degrees of freedom (≈ sample sizes), two top rows with p-values (areas under the curve), and, furthermore, the t-table is full of t-values, that, with ∞ degrees of freedom, become equal to z-values.

## 7 Conclusion

We conclude that basic knowledge of logarithms is convenient for a better understanding of many statistical methods. Odds ratio tests (Chap. 44), log likelihood ratio tests (Chap. 46), Markov modeling (Chap. 55), and many regression models use logarithmic transformations.

## 8 Note

More background, theoretical and mathematical information of logarithmic transformations is given in the Chaps. 28, 30, 44, 45, 46, 47, 48, 49, and 55.

# Chapter 44
# Odds Ratios, a Short-Cut for Analyzing Cross-Tabs

## 1 General Purpose

The odds ratio test is just like the chi-square test (Chap. 38) applicable for testing cross-tabs.

The advantage of the odds ratio test is, that an odds ratio value can be calculated. The odds ratio value is, just like the relative risk, an estimate of the chance of having had an event in group 1 as compared to that of having had it in group 2. An odds ratio value of 1.000 indicates no difference between the two groups. This chapter gives examples of odds ratio analyses.

## 2 Schematic Overview of Type of Data File

| Outcome binary | predictor binary |
|---|---|
| . | . |
| . | . |
| . | . |
| . | . |
| . | . |
| . | . |
| . | . |
| . | . |
| . | . |
| | |
| | |
| | |
| | |
| | |
| | |



T.J. Cleophas, A.H. Zwinderman, *Clinical Data Analysis on a Pocket Calculator*,
DOI 10.1007/978-3-319-27104-0_44

## 3 Primary Scientific Question

How can the odds ratio method be applied for testing interaction in a 2 × 2 contingency table, otherwise called 2 × 2 cross-tab, and for estimating relative risks of an event.

## 4 Example 1

| | events | no events | |
|---|---|---|---|
| | numbers of patients | | |
| group 1 | 15(a) | 20(b) | 35(a + b) |
| group 2 | 15(c) | 5(d) | 20(c + d) |
| | 30(a + c) | 25(b + d) | 55(a + b + c + d) |

-------------------------------

The odds of an event = the number of patients in a group with an event divided by the number without. In group 1 the odds of an event equals = a/b.

The odds ratio (OR) of group 1 compared to group 2

= (a/b)/(c/d)
= (15/20)/(15/5)
= 0.25

lnOR = ln 0.25 = −1.386 (ln = natural logarithm)

The standard error (SE) of the above term

= √ (1/a + 1/b + 1/c + 1/d)
= √ (1/15 + 1/20 + 1/15 + 1/5)
= √ 0.38333
= 0.619

The odds ratio can be tested using the z-test (Chap. 36).

The test-statistic = z-value
= (lnOR)/SE
= −1.386/0.619
= −2.239

If this value is smaller than −2 or larger than +2 (or more precisely −1.96 and +1.96), then the odds ratio is significantly different from 1 with two-tail p-value of <0.05, as shown in the bottom row of the t-table of the next page. An odds ratio of 1 means that there is no difference in events between group 1 and group 2.

| df | One-Tail = .4<br>Two-Tail = .8 | .25<br>.5 | .1<br>.2 | .05<br>.1 | .025<br>.05 | .01<br>.02 | .005<br>.01 | .0025<br>.005 | .001<br>.002 | .0005<br>.001 |
|---|---|---|---|---|---|---|---|---|---|---|
| 1 | 0.325 | 1.000 | 3.078 | 6.314 | 12.706 | 31.821 | 63.657 | 127.32 | 318.31 | 636.62 |
| 2 | 0.289 | 0.816 | 1.886 | 2.920 | 4.303 | 6.965 | 9.925 | 14.089 | 22.327 | 31.598 |
| 3 | 0.277 | 0.765 | 1.638 | 2.353 | 3.182 | 4.541 | 5.841 | 7.453 | 10.214 | 12.924 |
| 4 | 0.271 | 0.741 | 1.533 | 2.132 | 2.776 | 3.747 | 4.604 | 5.598 | 7.173 | 8.610 |
| 5 | 0.267 | 0.727 | 1.476 | 2.015 | 2.571 | 3.365 | 4.032 | 4.773 | 5.893 | 6.869 |
| 6 | 0.265 | 0.718 | 1.440 | 1.943 | 2.447 | 3.143 | 3.707 | 4.317 | 5.208 | 5.959 |
| 7 | 0.263 | 0.711 | 1.415 | 1.895 | 2.365 | 2.998 | 3.499 | 4.029 | 4.785 | 5.408 |
| 8 | 0.262 | 0.706 | 1.397 | 1.860 | 2.306 | 2.896 | 3.355 | 3.833 | 4.501 | 5.041 |
| 9 | 0.261 | 0.703 | 1.383 | 1.833 | 2.262 | 2.821 | 3.250 | 3.690 | 4.297 | 4.781 |
| 10 | 0.260 | 0.700 | 1.372 | 1.812 | 2.228 | 2.764 | 3.169 | 3.581 | 4.144 | 4.587 |
| 11 | 0.260 | 0.697 | 1.363 | 1.796 | 2.201 | 2.718 | 3.106 | 3.497 | 4.025 | 4.437 |
| 12 | 0.259 | 0.695 | 1.356 | 1.782 | 2.179 | 2.681 | 3.055 | 3.428 | 3.930 | 4.318 |
| 13 | 0.259 | 0.694 | 1.350 | 1.771 | 2.160 | 2.650 | 3.012 | 3.372 | 3.852 | 4.221 |
| 14 | 0.258 | 0.692 | 1.345 | 1.761 | 2.145 | 2.624 | 2.977 | 3.326 | 3.787 | 4.140 |
| 15 | 0.258 | 0.691 | 1.341 | 1.753 | 2.131 | 2.602 | 2.947 | 3.286 | 3.733 | 4.073 |
| 16 | 0.258 | 0.690 | 1.337 | 1.746 | 2.120 | 2.583 | 2.921 | 3.252 | 3.686 | 4.015 |
| 17 | 0.257 | 0.689 | 1.333 | 1.740 | 2.110 | 2.567 | 2.898 | 3.222 | 3.646 | 3.965 |
| 18 | 0.257 | 0.688 | 1.330 | 1.734 | 2.101 | 2.552 | 2.878 | 3.197 | 3.610 | 3.922 |
| 19 | 0.257 | 0.688 | 1.328 | 1.729 | 2.093 | 2.539 | 2.861 | 3.174 | 3.579 | 3.883 |
| 20 | 0.257 | 0.687 | 1.325 | 1.725 | 2.086 | 2.528 | 2.845 | 3.153 | 3.552 | 3.850 |
| 21 | 0.257 | 0.686 | 1.323 | 1.721 | 2.080 | 2.518 | 2.831 | 3.135 | 3.527 | 3.819 |
| 22 | 0.256 | 0.686 | 1.321 | 1.717 | 2.074 | 2.508 | 2.819 | 3.119 | 3.505 | 3.792 |
| 23 | 0.256 | 0.685 | 1.319 | 1.714 | 2.069 | 2.500 | 2.807 | 3.104 | 3.485 | 3.767 |
| 24 | 0.256 | 0.685 | 1.318 | 1.711 | 2.064 | 2.492 | 2.797 | 3.091 | 3.467 | 3.745 |
| 25 | 0.256 | 0.684 | 1.316 | 1.708 | 2.060 | 2.485 | 2.787 | 3.078 | 3.450 | 3.725 |
| 26 | 0.256 | 0.684 | 1.315 | 1.706 | 2.056 | 2.479 | 2.779 | 3.067 | 3.435 | 3.707 |
| 27 | 0.256 | 0.684 | 1.314 | 1.703 | 2.052 | 2.473 | 2.771 | 3.057 | 3.421 | 3.690 |
| 28 | 0.256 | 0.683 | 1.313 | 1.701 | 2.048 | 2.467 | 2.763 | 3.047 | 3.408 | 3.674 |
| 29 | 0.256 | 0.683 | 1.311 | 1.699 | 2.045 | 2.462 | 2.756 | 3.038 | 3.396 | 3.659 |
| 30 | 0.256 | 0.683 | 1.310 | 1.697 | 2.042 | 2.457 | 2.750 | 3.030 | 3.385 | 3.646 |
| 40 | 0.255 | 0.681 | 1.303 | 1.684 | 2.021 | 2.423 | 2.704 | 2.971 | 3.307 | 3.551 |
| 60 | 0.254 | 0.679 | 1.296 | 1.671 | 2.000 | 2.390 | 2.660 | 2.915 | 3.232 | 3.460 |
| 120 | 0.254 | 0.677 | 1.289 | 1.658 | 1.980 | 2.358 | 2.617 | 2.860 | 3.160 | 3.373 |
| ∞ | 0.253 | 0.674 | 1.282 | 1.645 | 1.960 | 2.326 | 2.576 | 2.807 | 3.090 | 3.291 |

The left-end column of the above t-table gives degrees of freedom ($\approx$ sample sizes), two top rows with p-values (areas under the curve), and furthermore full of t-values, that, with $\infty$ degrees of freedom, are equal to z-values.

## 5 Example 2

| | events | no events | |
|---|---|---|---|
| | number of patients | | |
| group 1 | 16(a) | 26(b) | 42(a + b) |
| group 2 | 5(c) | 30(d) | 35(c + d) |
| | 21(a + c) | 56(b + d) | 77(a + b + c + d) |

Test with OR whether there is a significant difference between group 1 and 2. See for procedure also example 1.

OR = (16/26)/(5/30)
= 3.69

lnOR = 1.3056 (ln = natural logarithm see the above example)

SE = √ (1/16 + 1/26 + 1/5 + 1/30)
= √ 0.334333
= 0.578

z-value = 1.3056/0.578
= 2.259

Because this value is larger than 2, a two-tail p-value of <0.05 is observed, 0.024 to be precise (numerous "p-calculator for z-values" sites in Google will help you calculate a more precise p-value if required. However, the above t-table may be sufficiently precise for your purposes.

## 6 Conclusion

The odds ratio test is, like the z-test (Chap. 36), and the chi-square test (Chap. 38) applicable for testing cross-tabs. A nice thing about odds ratio tests is that, unlike the other tests, they provide odds ratios, that can be interpreted as relative risks. E. g., an odds ratio of 3.69 means that one group performs about 3.7 better than the other. Odds ratio methods are easy and fast, and kind of short -cut for analyzing cross-tabs.

## 7 Note

More background, theoretical and mathematical information of odds ratio tests is given in Statistics applied to clinical studies 5th edition, Chap. 3, Springer Heidelberg Germany, 2012, from the same authors.

# Chapter 45
# Logodds, the Basis of Logistic Regression

## 1 General Purpose

Logistic regression is much similar to linear regression (see Chap. 8). The difference is the type of outcome variable, which is continuous with linear regression and binary with logistic regression. In order for logistic regression to work, we need to transform the binary outcome into the odds of responding, or rather the logodds of responding. In a population

$$\text{the odds of an infarction} = \frac{\text{the number of patients with infarct}}{\text{number of patients without.}}$$

The easiest way to understand the term odds is to think of it as though it is the risk.

The odds or risk of an infarction is correlated with age: the older, the larger the odds.

Now how does it correlate with age? As shown underneath it is not at all linear.

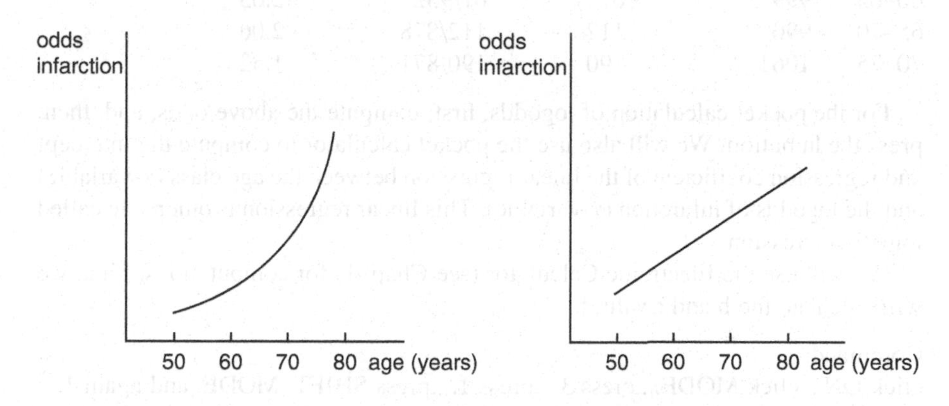


However, if we transform the underneath linear model



T.J. Cleophas, A.H. Zwinderman, *Clinical Data Analysis on a Pocket Calculator*,
DOI 10.1007/978-3-319-27104-0_45

$$y = a + bx$$

into a loglinear model

$$\ln \text{odds} = a + b \times (x = \text{age}),$$

then, all of a sudden, we will observe a close to linear relationship (the above right graph). This present from heaven can be used for statistical testing. The current chapter assesses how ln odds, often called logodds, can be used for testing studies with binary outcome data, like numbers of responders to treatment yes or no.

## 2 Primary Scientific Question

How can we use logodds for modeling binary outcome data for the purpose of making predictions from them.

## 3 Data Example

In a random population of subjects between 50 and 75, we register 5 year age classes for each subject and follow them for 3 years. After that the numbers of infarctions are counted per age class.

| ageclass | population size | infarctions | odds of infarct | logodds |
|---|---|---|---|---|
| 50–55 | 977 | 8 | 8/969 | −4.80 |
| 55–60 | 1010 | 42 | 42/968 | −3.14 |
| 60–65 | 999 | 67 | 67/932 | −2.63 |
| 65–70 | 990 | 112 | 112/878 | −2.06 |
| 70–75 | 1061 | 190 | 190/871 | −1.52 |

For the pocket calculation of logodds, first, compute the above odds, and, then, press the ln button. We will also use the pocket calculator to compute the intercept and regression coefficient of the linear regression between the age class (x-variable) and the logodds of infarction (y-variable). This linear regression is otherwise called logistic regression.

We will use the Electronic Calculator (see Chap. 1) for computations. First, we will calculate the b and r values.

Command:
click ON....click MODE....press 3....press 1....press SHIFT, MODE, and again 1.... press = ....start entering the data.... [55, −4.80]....[60, −3.14]....[65, −2.63]... [70. −2.06]...[75, −1.52]

In order to obtain the a value, press: shift, S-VAR, ▶, ▶, 1, = .
In order to obtain the b value, press: shift, S-VAR, ▶, ▶, 2, = .

The underneath values are obtained.

$a = -12.8$
$b = 0.15$

logodds = a + b (age class)

Now, we can use the above equation for making predictions about the risk of having an infarction in the upcoming 3 years in future subjects.

For someone 50–55 years the logodds of infarction $= -12.8 + 0.15\ (55)$
$= -4.55$
the odds of infarction $= 0.0106 = 1.06\ \%$

For someone 70–75 years the logodds of infarction $= -12.8 + 0.15\ (75)$
$= -1.55$
the odds of infarction $= 0.212 = 21.2\ \%$

The risk is pretty much similar to the odds, particularly, with small risks, and is equal to risk = 1/(1 + 1/odds)

e.g., with odds $= 21.2\ \%$
risk $= 1/(1 + 1/0.212)$
$= 0.175 = 17.5\ \%$.

# 4 Conclusion

Logistic regression is not in pocket calculators, but can be used even so if you transform your outcome odds values into logodds values. We should add that logistic regression is a magnificent methodology with plenty applications, most of whom require the use of advanced statistical software. For example it can be used not only for predictive models of age and infarction, but also for predictive models with multiple predictors like risk factors and patient characteristics for risk management assessments. It also can be used for efficacy analysis of survival studies and exploratory purposes. More information is given in the Chaps. 17, 19, 21, 49, and 65, in Statistics applied to clinical studies 5th edition, Springer Heidelberg Germany, 2012, from the same authors.

## 5 Note

More background, theoretical and mathematical information of logistic regression is given in SPSS for starters and 2nd levelers 2nd edition, Chap. 36, 37, 38, and 39, Springer Heidelberg Germany, 2015, from the same authors.

# Chapter 46
# Log Likelihood Ratio Tests for the Best Precision

## 1 General Purpose

The sensitivity of the chi-square test (Chap. 38), and the odds ratio test (Chap. 44) for testing cross-tabs is limited, and, not entirely, accurate, if the values in one or more cells is smaller than 5. The log likelihood ratio test may be an adequate alternative with generally better sensitivity, and, so, it must be absolutely considered. This chapter assesses how it works.

## 2 Schematic Overview of Type of Data File

| Outcome |
|---|
| . |
| . |
| . |
| . |
| . |
| . |
| . |
| . |
| . |
| |
| |
| |



T.J. Cleophas, A.H. Zwinderman, *Clinical Data Analysis on a Pocket Calculator*,
DOI 10.1007/978-3-319-27104-0_46

## 3 Primary Scientific Question

Is the log likelihood ratio test adequate and more sensitive for testing 2 × 2 contingency table (cross-tabs, 2 × 2 interaction tables) compared to standard methods.

## 4 Example 1

A group of citizens is taking a pharmaceutical company to court for misrepresenting the danger of fatal rhabdomyolysis due to statin treatment.

| | Patients with rhabdomyolysis | patients without |
|---|---|---|
| company | 1 (a) | 309999 (b) |
| citizens | 4 (c) | 300289 (d) |

$p_{co}$ = proportion given by the pharmaceutical company = a/(a + b) = 1/310000
$p_{ci}$ = proportion given by the citizens = c/(c + d) = 4/300293
We make use of the z-test (Chap. 36) for testing log likelihood ratios.

As it can be shown that −2 log likelihood ratio equals $z^2$, we can test the significance of difference between the two proportions.

$$\text{Log likelihood ratio} = 4\log\frac{1/310000}{4/300293} + 300289\ \log\frac{1-1/310000}{1-4/300293}$$
$$= -2.641199$$
$$-2\ \text{log likelihood ratio} = -2 \times -2.641199$$
$$= 5.2824\ (p < 0.05,\ \text{because}\ z > 2).$$
$$= z^2$$

A z – value larger than 2 (actually 1.960, see bottom row of underneath t-table for all z-values) means a significant difference in your data. Here the z-value equals √ 5.2824 = 2.29834. The "p-calculator for z-values" in Google tells you that a more precise p – value = 0.0215, anyway much smaller than 0.05.

We should note here that both the odds ratio test and chi-square test produced a non-significant result of these data (p > 0.05). Indeed, the log likelihood ratio test is much more sensitive than the other tests for the same data, which might once in a while be a blessing for desperate investigators.

| *df* | One-Tail = .4<br>Two-Tail = .8 | .25<br>.5 | .1<br>.2 | .05<br>.1 | .025<br>.05 | .01<br>.02 | .005<br>.01 | .0025<br>.005 | .001<br>.002 | .0005<br>.001 |
|---|---|---|---|---|---|---|---|---|---|---|
| 1 | 0.325 | 1.000 | 3.078 | 6.314 | 12.706 | 31.821 | 63.657 | 127.32 | 318.31 | 636.62 |
| 2 | 0.289 | 0.816 | 1.886 | 2.920 | 4.303 | 6.965 | 9.925 | 14.089 | 22.327 | 31.598 |
| 3 | 0.277 | 0.765 | 1.638 | 2.353 | 3.182 | 4.541 | 5.841 | 7.453 | 10.214 | 12.924 |
| 4 | 0.271 | 0.741 | 1.533 | 2.132 | 2.776 | 3.747 | 4.604 | 5.598 | 7.173 | 8.610 |
| 5 | 0.267 | 0.727 | 1.476 | 2.015 | 2.571 | 3.365 | 4.032 | 4.773 | 5.893 | 6.869 |
| 6 | 0.265 | 0.718 | 1.440 | 1.943 | 2.447 | 3.143 | 3.707 | 4.317 | 5.208 | 5.959 |
| 7 | 0.263 | 0.711 | 1.415 | 1.895 | 2.365 | 2.998 | 3.499 | 4.029 | 4.785 | 5.408 |
| 8 | 0.262 | 0.706 | 1.397 | 1.860 | 2.306 | 2.896 | 3.355 | 3.833 | 4.501 | 5.041 |
| 9 | 0.261 | 0.703 | 1.383 | 1.833 | 2.262 | 2.821 | 3.250 | 3.690 | 4.297 | 4.781 |
| 10 | 0.260 | 0.700 | 1.372 | 1.812 | 2.228 | 2.764 | 3.169 | 3.581 | 4.144 | 4.587 |
| 11 | 0.260 | 0.697 | 1.363 | 1.796 | 2.201 | 2.718 | 3.106 | 3.497 | 4.025 | 4.437 |
| 12 | 0.259 | 0.695 | 1.356 | 1.782 | 2.179 | 2.681 | 3.055 | 3.428 | 3.930 | 4.318 |
| 13 | 0.259 | 0.694 | 1.350 | 1.771 | 2.160 | 2.650 | 3.012 | 3.372 | 3.852 | 4.221 |
| 14 | 0.258 | 0.692 | 1.345 | 1.761 | 2.145 | 2.624 | 2.977 | 3.326 | 3.787 | 4.140 |
| 15 | 0.258 | 0.691 | 1.341 | 1.753 | 2.131 | 2.602 | 2.947 | 3.286 | 3.733 | 4.073 |
| 16 | 0.258 | 0.690 | 1.337 | 1.746 | 2.120 | 2.583 | 2.921 | 3.252 | 3.686 | 4.015 |
| 17 | 0.257 | 0.689 | 1.333 | 1.740 | 2.110 | 2.567 | 2.898 | 3.222 | 3.646 | 3.965 |
| 18 | 0.257 | 0.688 | 1.330 | 1.734 | 2.101 | 2.552 | 2.878 | 3.197 | 3.610 | 3.922 |
| 19 | 0.257 | 0.688 | 1.328 | 1.729 | 2.093 | 2.539 | 2.861 | 3.174 | 3.579 | 3.883 |
| 20 | 0.257 | 0.687 | 1.325 | 1.725 | 2.086 | 2.528 | 2.845 | 3.153 | 3.552 | 3.850 |
| 21 | 0.257 | 0.686 | 1.323 | 1.721 | 2.080 | 2.518 | 2.831 | 3.135 | 3.527 | 3.819 |
| 22 | 0.256 | 0.686 | 1.321 | 1.717 | 2.074 | 2.508 | 2.819 | 3.119 | 3.505 | 3.792 |
| 23 | 0.256 | 0.685 | 1.319 | 1.714 | 2.069 | 2.500 | 2.807 | 3.104 | 3.485 | 3.767 |
| 24 | 0.256 | 0.685 | 1.318 | 1.711 | 2.064 | 2.492 | 2.797 | 3.091 | 3.467 | 3.745 |
| 25 | 0.256 | 0.684 | 1.316 | 1.708 | 2.060 | 2.485 | 2.787 | 3.078 | 3.450 | 3.725 |
| 26 | 0.256 | 0.684 | 1.315 | 1.706 | 2.056 | 2.479 | 2.779 | 3.067 | 3.435 | 3.707 |
| 27 | 0.256 | 0.684 | 1.314 | 1.703 | 2.052 | 2.473 | 2.771 | 3.057 | 3.421 | 3.690 |
| 28 | 0.256 | 0.683 | 1.313 | 1.701 | 2.048 | 2.467 | 2.763 | 3.047 | 3.408 | 3.674 |
| 29 | 0.256 | 0.683 | 1.311 | 1.699 | 2.045 | 2.462 | 2.756 | 3.038 | 3.396 | 3.659 |
| 30 | 0.256 | 0.683 | 1.310 | 1.697 | 2.042 | 2.457 | 2.750 | 3.030 | 3.385 | 3.646 |
| 40 | 0.255 | 0.681 | 1.303 | 1.684 | 2.021 | 2.423 | 2.704 | 2.971 | 3.307 | 3.551 |
| 60 | 0.254 | 0.679 | 1.296 | 1.671 | 2.000 | 2.390 | 2.660 | 2.915 | 3.232 | 3.460 |
| 120 | 0.254 | 0.677 | 1.289 | 1.658 | 1.980 | 2.358 | 2.617 | 2.860 | 3.160 | 3.373 |
| ∞ | 0.253 | 0.674 | 1.282 | 1.645 | 1.960 | 2.326 | 2.576 | 2.807 | 3.090 | 3.291 |

The above t-table gives in the left-end column degrees of freedom (df), ($\approx$ sample sizes), two top rows with p-values (areas under the curve), and furthermore plenty t-values, that, with $\infty$ degrees of freedom, have become equal to z-values. We should emphasize, that, instead of the above t-table, the one-degree-of-freedom-row of the chi-square table can be used, because this row produces $z^2$ – values.

The underneath chi-square table has an upper row with areas under the curve, a left-end column with degrees of freedom, and a whole lot of chi-square values. The one-degree-of-freedom-row of the chi-square table has values equal to the squared values of the bottom row of the t-table.

Chi-squared distribution

| | Two-tailed *P*-value | | | |
|---|---|---|---|---|
| *df* | 0.10 | 0.05 | 0.01 | 0.001 |
| 1 | 2.706 | 3.841 | 6.635 | 10.827 |
| 2 | 4.605 | 5.991 | 9.210 | 13.815 |
| 3 | 6.251 | 7.851 | 11.345 | 16.266 |
| 4 | 7.779 | 9.488 | 13.277 | 18.466 |
| 5 | 9.236 | 11.070 | 15.086 | 20.515 |
| 6 | 10.645 | 12.592 | 16.812 | 22.457 |
| 7 | 12.017 | 14.067 | 18.475 | 24.321 |
| 8 | 13.362 | 15.507 | 20.090 | 26.124 |
| 9 | 14.684 | 16.919 | 21.666 | 27.877 |
| 10 | 15.987 | 18.307 | 23.209 | 29.588 |
| 11 | 17.275 | 19.675 | 24.725 | 31.264 |
| 12 | 18.549 | 21.026 | 26.217 | 32.909 |
| 13 | 19.812 | 22.362 | 27.688 | 34.527 |
| 14 | 21.064 | 23.685 | 29.141 | 36.124 |
| 15 | 22.307 | 24.996 | 30.578 | 37.698 |
| 16 | 23.542 | 26.296 | 32.000 | 39.252 |
| 17 | 24.769 | 27.587 | 33.409 | 40.791 |
| 18 | 25.989 | 28.869 | 34.805 | 42.312 |
| 19 | 27.204 | 30.144 | 36.191 | 43.819 |
| 20 | 28.412 | 31.410 | 37.566 | 45.314 |
| 21 | 29.615 | 32.671 | 38.932 | 46.796 |
| 22 | 30.813 | 33.924 | 40.289 | 48.268 |
| 23 | 32.007 | 35.172 | 41.638 | 49.728 |
| 24 | 33.196 | 36.415 | 42.980 | 51.179 |
| 25 | 34.382 | 37.652 | 44.314 | 52.619 |
| 26 | 35.536 | 38.885 | 45.642 | 54.051 |
| 27 | 36.741 | 40.113 | 46.963 | 55.475 |
| 28 | 37.916 | 41.337 | 48.278 | 56.892 |
| 29 | 39.087 | 42.557 | 49.588 | 58.301 |
| 30 | 40.256 | 43.773 | 50.892 | 59.702 |
| 40 | 51.805 | 55.758 | 63.691 | 73.403 |
| 50 | 63.167 | 67.505 | 76.154 | 86.660 |
| 60 | 74.397 | 79.082 | 88.379 | 99.608 |
| 70 | 85.527 | 90.531 | 100.43 | 112.32 |
| 80 | 96.578 | 101.88 | 112.33 | 124.84 |
| 90 | 107.57 | 113.15 | 124.12 | 137.21 |
| 100 | 118.50 | 124.34 | 135.81 | 149.45 |

## 5 Example 2

Two group of 15 patients at risk for arrhythmias were assessed for the development of torsade de points after calcium channel blockers treatment

| | Patients with torsade de points | patients without |
|---|---|---|
| Calcium channel blocker 1 | 5 | 10 |
| Calcium channel blocker 2 | 9 | 6 |

The proportion of patients with event from calcium channel blocker 1 is 5/15, from blocker 2 it is 9/15.

Log likelihood ratio $= 9 \log \frac{5/15}{9/15} + 6 \log \frac{1-5/15}{1-9/15}$

$= -2.25$

−2 log likelihood ratio $= 4.50$

$= z^2$

z – value $= \sqrt{4.50} = 2.1213$

p – value $<0.05$, because $z > 2$.

The traditional chi-square test of these data was non-significant ($p > 0.05$), (Chap. 38). You can check for yourself, that, with the odds ratio test (Chap. 44), this will be equally so.

## 6 Example 3

Two groups of patients with stage IV New York Heart Association heart failure were assessed for clinical admission while on two beta-blockers.

| | Patients with clinical admission | patients without |
|---|---|---|
| Beta blocker 1 | 77 | 62 |
| Beta blocker 2 | 103 | 46 |

The proportion of patients with event while on beta blocker 1 is 77/139, while on beta blocker 2 it is 103/149.

Log likelihood ratio $= 103 \log \frac{77/139}{103/149} + 46 \log \frac{1-77/139}{1-103/149}$

$= -5.882$

−2 log likelihood ratio $= 11.766$

$= z^2$

z – value $= \sqrt{11.766} = 3.43016$

p – value $<0.002$, because $z > 3.090$ (see the above t-table).

Both the odds ratio test and chi-square test were also significant. However, at lower levels of significance, both p-values $0.01 < p < 0.05$.

## 7 Conclusion

The sensitivity of the traditional tests for testing cross-tabs is limited, and, not entirely, accurate, with cells smaller than 5. The log likelihood ratio test is an adequate alternative with generally better sensitivity. It is lovely to use it, if your traditional test can not reject the null hypothesis of your study with a p-value a bit larger than 0.05. Your chance is big, that the current test will produce a p-value just under 0.05.

## 8 Note

More background, theoretical and mathematical information of log likelihood ratio tests is given in Statistics applied to clinical studies 5th edition, Chap. 4, Springer Heidelberg Germany, 2012, from the same authors.

# Chapter 47
# Hierarchical Loglinear Models for Higher Order Cross-Tabs

## 1 General Purpose

The Pearson chi-square test is traditionally used for analyzing two dimensional contingency tables, otherwise called cross-tabs or interaction matrices (Chap. 38). It can anwer questions like: is the risk of falling out of bed different between the departments of surgery and internal medicine (Chaps. 37 and 38). The analysis is, however, very limited, because only the interaction between the two variables, e.g., (1) falling out of bed (yes, no) and (2) department (one or the other) is assessed. In contrast, in an observational data set we may be interested in the effects of the two variables separately:

1. is there a significant difference between the numbers of patients falling out of bed and the patients who don't (the main effect of variable 1),
2. is there a difference between the numbers of patients being in one department and those being in the other (the main effect of variable 2).

The Pearson test is unable to answer such questions. Hierarchical loglinear modeling is a pretty novel methodology adequate for the purpose, but not yet widely available. In SPSS versions 16–23 it is not in the menu, but only accessible through syntax commands.

In order to simultaneously analyze, in a 2 × 2 cross tab, the effects of the main variable in addtion to their interaction, ANOVA (analysis of variance) might be considered. In ANOVA with two predictor factors and one outcome, outcome observations are often modeled as a linear combination of:

1 the grand mean
2 the main effect of the first predictor
3 the main effect of the second predictor
4 the interaction effect of the first and the second predictor.



T.J. Cleophas, A.H. Zwinderman, *Clinical Data Analysis on a Pocket Calculator*,
DOI 10.1007/978-3-319-27104-0_47

However, ANOVA requires continuous outcome variables and contingency tables consist of counted data (numbers of responders, numbers of yes answers), like numbers of patients falling out of bed. With cell-counts data, like interaction matrices, traditional ANOVA is impossible, because the outcome-observations must be modeled as the product of the above 4 effects, rather than their linear add-up sum. The trick is to transform the multiplicative model into a linear model using

logarithmic transformation (ln = natural logarithm is always used).

Outcome = 1*2*3*4 (* = symbol of multiplication)
Log outcome = log 1 + log 2 + log 3 + log 4

## 2 Schematic Overview of Type of Data File

| Treatment modality (1 and 2) | Outcome (1 and 2) |
|---|---|
| . | . |
| . | . |
| . | . |
| . | . |
| . | . |
| . | . |
| . | . |
| | |
| | |

## 3 Primary Scientific Question

Can hierarchical loglinear modeling simultaneously assess the effects of the main variables in addition to their interaction?

# 4 Data Example

A simple 2 × 2 contingency table is given with two treatment groups as row variable and the presence of sleeplessness as column variable. A loglinear analysis is given underneath. Loglikelihood ratio tests are used for the computations (see also the Chap. 46).

| | column | 1 | 2 | |
|---|---|---|---|---|
| row | 1 | 50 | 150 | 200 |
| | 2 | 90 | 60 | 150 |
| | | 140 | 210 | 350 |

All counts have to be logarithmically transformed (ln 50 = 3.912 etc.).

| | column | 1 | 2 | |
|---|---|---|---|---|
| row | 1 | 3.912 | 5.011 | 5.298 |
| | 2 | 4.500 | 4.049 | 5.011 |
| | | 4.942 | 5.347 | 5.848 |

## 4.1 *First Order Effects*

Is there a significant main effect of the column variable (is the number of sleepy people significantly different from that of non-sleepy people). Expected log frequencies log(350/2) = 5.165. The loglikelihood ratio (LLR) chi-square test is used for testing.

In this test −2 loglikelihood ratio should be larger than $z^2 = 2^2 = 4$ in order to obtain statistical significance at a $p < 0.05$ level, and $z^2$ = one-degree-of-freedom chi-square value as explained in the Chap. 46).

$$-2\ LLR_{column} = 2 * (140*(4.942-5.165) + 210*(5.347-5.165))$$
$$= 140.0,$$

* = symbol of multiplication.

The underneath chi-square table has an upper row with areas under the curve (p-values), a left-end column with degrees of freedom (df), and, furthermore, a whole lot of chi-square values.

Chi-squared distribution

| | Two-tailed *P*-value | | | |
|---|---|---|---|---|
| *df* | 0.10 | 0.05 | 0.01 | 0.001 |
| 1 | 2.706 | 3.841 | 6.635 | 10.827 |
| 2 | 4.605 | 5.991 | 9.210 | 13.815 |
| 3 | 6.251 | 7.851 | 11.345 | 16.266 |
| 4 | 7.779 | 9.488 | 13.277 | 18.466 |
| 5 | 9.236 | 11.070 | 15.086 | 20.515 |
| 6 | 10.645 | 12.592 | 16.812 | 22.457 |
| 7 | 12.017 | 14.067 | 18.475 | 24.321 |
| 8 | 13.362 | 15.507 | 20.090 | 26.124 |
| 9 | 14.684 | 16.919 | 21.666 | 27.877 |
| 10 | 15.987 | 18.307 | 23.209 | 29.588 |
| 11 | 17.275 | 19.675 | 24.725 | 31.264 |
| 12 | 18.549 | 21.026 | 26.217 | 32.909 |
| 13 | 19.812 | 22.362 | 27.688 | 34.527 |
| 14 | 21.064 | 23.685 | 29.141 | 36.124 |
| 15 | 22.307 | 24.996 | 30.578 | 37.698 |
| 16 | 23.542 | 26.296 | 32.000 | 39.252 |
| 17 | 24.769 | 27.587 | 33.409 | 40.791 |
| 18 | 25.989 | 28.869 | 34.805 | 42.312 |
| 19 | 27.204 | 30.144 | 36.191 | 43.819 |
| 20 | 28.412 | 31.410 | 37.566 | 45.314 |
| 21 | 29.615 | 32.671 | 38.932 | 46.796 |
| 22 | 30.813 | 33.924 | 40.289 | 48.268 |
| 23 | 32.007 | 35.172 | 41.638 | 49.728 |
| 24 | 33.196 | 36.415 | 42.980 | 51.179 |
| 25 | 34.382 | 37.652 | 44.314 | 52.619 |
| 26 | 35.536 | 38.885 | 45.642 | 54.051 |
| 27 | 36.741 | 40.113 | 46.963 | 55.475 |
| 28 | 37.916 | 41.337 | 48.278 | 56.892 |
| 29 | 39.087 | 42.557 | 49.588 | 58.301 |
| 30 | 40.256 | 43.773 | 50.892 | 59.702 |
| 40 | 51.805 | 55.758 | 63.691 | 73.403 |
| 50 | 63.167 | 67.505 | 76.154 | 86.660 |
| 60 | 74.397 | 79.082 | 88.379 | 99.608 |
| 70 | 85.527 | 90.531 | 100.43 | 112.32 |
| 80 | 96.578 | 101.88 | 112.33 | 124.84 |
| 90 | 107.57 | 113.15 | 124.12 | 137.21 |
| 100 | 118.50 | 124.34 | 135.81 | 149.45 |

A chi-square value of 140, one degree of freedom, means that p is much $<0.001$.

Is there a significant main effect of the row variable (is the numbers of treatments in group 1 significantly different from that of group 2). Expected log frequencies = log (350/2) = 5.165.

$$-2\ LLR_{row} = 2 * (200*(5.298-5.165) + 150*(5.011-5.165)) = 7.0,$$

A chi-square value of 7.0 and 1df, means that $p < 0.01$.

## 4.2 *Second Order Effects*

Is there a significant interaction between the row and column variable. The loglikelihood ratio (LLR) chi-square test is again used for testing.

$$-2\ LLR_{column\ x\ row} = 2 * [(200*(5.298-5.165) + 150*(5.011-5.165) + 140*(4.942-5.165) + 210*(5.347-5.165)] = 21.0,$$

1 df, $p < 0.001$.

The traditional Pearson chi-square test for "row x column" is similarly very significant, although with a larger chi-square value. We will use the pocket calculator method (Chap. 38).

$$\text{Pearson chi-square}_{column\ x\ row} = [(50*60 - 90*150)^2 * 350]/(140*210*150*200) = 43.75,$$

1 df, $P < 0.0001$.

* = symbol of multiplication.

The Pearson chi-square value is larger than the above second order log likelihood ratio test. This is, because the former does not account first order effects. In general, if you account more, then you will prove less.

# 5 Conclusion

The above example shows that logarithmic transformation of multiplicative models for analyzing contingency tables can readily provide first and second statistics with the help of log likelihood ratio tests. Also, in practice, higher order contingency tables do exist. E.g, we may want to know, whether variables like ageclass, gender, and other patient characteristics interact with the variables (1) and (2). Calculations are, of course, increasingly complex, and a pocket calculator assessment is impossible. In Chap. 52, SPSS for starters and second levelers 2nd edition, Springer

Heidelberg Germany, 2015, from the same authors, examples of third, and fourth order hierarchical loglinear models are given.

## 6 Note

More background, theoretical and mathematical information of hierarchical log linear models is given in SPSS for starters and second levelers 2nd edition, Chap. 52, Springer Heidelberg Germany, 2015, from the same authors.

# Chapter 48
# McNemar's Tests for Paired Cross-Tabs

## 1 General Purpose

The Chaps. 36, 37, 38, 39, 44, and 46 have reviewed methods for analyzing cross-tabs of two groups of patients, otherwise called unpaired cross-tabs. Sometimes, a single group of subjects is assessed twice, and, then, we will obtain a cross-tab slightly different from the traditional unpaired cross-tabs. McNemar's test must be applied for analyzing these kinds of data.

## 2 Schematic Overview of Type of Data File

| Outcome-1 binary | outcome-2 binary |
|---|---|
| . | . |
| . | . |
| . | . |
| . | . |
| . | . |
| . | . |
| . | . |
| . | . |
| . | . |

 
T.J. Cleophas, A.H. Zwinderman, *Clinical Data Analysis on a Pocket Calculator*,
DOI 10.1007/978-3-319-27104-0_48

## 3 Primary Scientific Question

In a crossover study two diagnostic devices are assessed. We wish to know whether one performs better than the other.

## 4 Data Example, Chi-Square McNemar's Test

315 subjects are tested for hypertension using both an automated device (device-1) and a sphygmomanometer (device-2).

| | Device-1 | | |
|---|---|---|---|
| | + | − | total |
| Device-2 + | 184 | 54 | 238 |
| − | 14 | 63 | 77 |
| Total | 198 | 117 | 315 |

$$\text{Chi} - \text{square McNemar} = \frac{(54-14)^2}{54+14} = 23.5$$

184 subjects scored positive with both tests and 63 scored negative with both tests. These 247 subjects, therefore, give us no information about which of the two tests is more likely to score positive. The information we require is entirely contained in the 68 subjects for whom the tests did not agree (the discordant pairs).

The above equation shows how the chi-square value is calculated. The chi-square table (given underneath) finds the appropriate p-value. Like with $2 \times 2$ unpaired cross-tabs, we will have 1 degree of freedom (df). The 1 degree of freedom row of the chi-square table shows, that, with our result of 23.5 is a lot larger than 10.827, that the p-value will be a lot <0.001. We conclude, that the two devices produced significantly different results at $p < 0.001$.

Chi-squared distribution

| | Two-tailed *P*-value | | | |
|---|---|---|---|---|
| *df* | 0.10 | 0.05 | 0.01 | 0.001 |
| 1 | 2.706 | 3.841 | 6.635 | 10.827 |
| 2 | 4.605 | 5.991 | 9.210 | 13.815 |
| 3 | 6.251 | 7.851 | 11.345 | 16.266 |
| 4 | 7.779 | 9.488 | 13.277 | 18.466 |
| 5 | 9.236 | 11.070 | 15.086 | 20.515 |
| 6 | 10.645 | 12.592 | 16.812 | 22.457 |
| 7 | 12.017 | 14.067 | 18.475 | 24.321 |

(continued)

| | Two-tailed *P*-value | | | |
|---|---|---|---|---|
| *df* | 0.10 | 0.05 | 0.01 | 0.001 |
| 8 | 13.362 | 15.507 | 20.090 | 26.124 |
| 9 | 14.684 | 16.919 | 21.666 | 27.877 |
| 10 | 15.987 | 18.307 | 23.209 | 29.588 |
| 11 | 17.275 | 19.675 | 24.725 | 31.264 |
| 12 | 18.549 | 21.026 | 26.217 | 32.909 |
| 13 | 19.812 | 22.362 | 27.688 | 34.527 |
| 14 | 21.064 | 23.685 | 29.141 | 36.124 |
| 15 | 22.307 | 24.996 | 30.578 | 37.698 |
| 16 | 23.542 | 26.296 | 32.000 | 39.252 |
| 17 | 24.769 | 27.587 | 33.409 | 40.791 |
| 18 | 25.989 | 28.869 | 34.805 | 42.312 |
| 19 | 27.204 | 30.144 | 36.191 | 43.819 |
| 20 | 28.412 | 31.410 | 37.566 | 45.314 |
| 21 | 29.615 | 32.671 | 38.932 | 46.796 |
| 22 | 30.813 | 33.924 | 40.289 | 48.268 |
| 23 | 32.007 | 35.172 | 41.638 | 49.728 |
| 24 | 33.196 | 36.415 | 42.980 | 51.179 |
| 25 | 34.382 | 37.652 | 44.314 | 52.619 |
| 26 | 35.536 | 38.885 | 45.642 | 54.051 |
| 27 | 36.741 | 40.113 | 46.963 | 55.475 |
| 28 | 37.916 | 41.337 | 48.278 | 56.892 |
| 29 | 39.087 | 42.557 | 49.588 | 58.301 |
| 30 | 40.256 | 43.773 | 50.892 | 59.702 |
| 40 | 51.805 | 55.758 | 63.691 | 73.403 |
| 50 | 63.167 | 67.505 | 76.154 | 86.660 |
| 60 | 74.397 | 79.082 | 88.379 | 99.608 |
| 70 | 85.527 | 90.531 | 100.43 | 112.32 |
| 80 | 96.578 | 101.88 | 112.33 | 124.84 |
| 90 | 107.57 | 113.15 | 124.12 | 137.21 |
| 100 | 118.50 | 124.34 | 135.81 | 149.45 |

The above chi-square table has an upper row with areas under the curve (p-values), a left-end column with degrees of freedom (df), and a whole lot of chi-square values.

# 5 Data Example, McNemar's Z-Test

Instead of the above chi-square McNemar's test, also a McNemar's z-test can be performed, and will produce identical results. Again 184 and 63 patients need not be taken into account. The test works as follows.

$$z = (54 - 14)/\sqrt{(54 + 14)} = 4.85$$

A z-value > 1.960 means, that a significant difference between the two tests exists at $p < 0.05$. The above z-value is a lot larger than 1.96. The underneath t-table (bottom row) must be used for computing the p-value more precisely. A z-value of 4.85 is larger than 3.291, and, thus, the two tests are significantly from one another at a two-tail p-value < 0.001.

| df | One-Tail = .4<br>Two-Tail = .8 | .25<br>.5 | .1<br>.2 | .05<br>.1 | .025<br>.05 | .01<br>.02 | .005<br>.01 | .0025<br>.005 | .001<br>.002 | .0005<br>.001 |
|---|---|---|---|---|---|---|---|---|---|---|
| 1 | 0.325 | 1.000 | 3.078 | 6.314 | 12.706 | 31.821 | 63.657 | 127.32 | 318.31 | 636.62 |
| 2 | 0.289 | 0.816 | 1.886 | 2.920 | 4.303 | 6.965 | 9.925 | 14.089 | 22.327 | 31.598 |
| 3 | 0.277 | 0.765 | 1.638 | 2.353 | 3.182 | 4.541 | 5.841 | 7.453 | 10.214 | 12.924 |
| 4 | 0.271 | 0.741 | 1.533 | 2.132 | 2.776 | 3.747 | 4.604 | 5.598 | 7.173 | 8.610 |
| 5 | 0.267 | 0.727 | 1.476 | 2.015 | 2.571 | 3.365 | 4.032 | 4.773 | 5.893 | 6.869 |
| 6 | 0.265 | 0.718 | 1.440 | 1.943 | 2.447 | 3.143 | 3.707 | 4.317 | 5.208 | 5.959 |
| 7 | 0.263 | 0.711 | 1.415 | 1.895 | 2.365 | 2.998 | 3.499 | 4.029 | 4.785 | 5.408 |
| 8 | 0.262 | 0.706 | 1.397 | 1.860 | 2.306 | 2.896 | 3.355 | 3.833 | 4.501 | 5.041 |
| 9 | 0.261 | 0.703 | 1.383 | 1.833 | 2.262 | 2.821 | 3.250 | 3.690 | 4.297 | 4.781 |
| 10 | 0.260 | 0.700 | 1.372 | 1.812 | 2.228 | 2.764 | 3.169 | 3.581 | 4.144 | 4.587 |
| 11 | 0.260 | 0.697 | 1.363 | 1.796 | 2.201 | 2.718 | 3.106 | 3.497 | 4.025 | 4.437 |
| 12 | 0.259 | 0.695 | 1.356 | 1.782 | 2.179 | 2.681 | 3.055 | 3.428 | 3.930 | 4.318 |
| 13 | 0.259 | 0.694 | 1.350 | 1.771 | 2.160 | 2.650 | 3.012 | 3.372 | 3.852 | 4.221 |
| 14 | 0.258 | 0.692 | 1.345 | 1.761 | 2.145 | 2.624 | 2.977 | 3.326 | 3.787 | 4.140 |
| 15 | 0.258 | 0.691 | 1.341 | 1.753 | 2.131 | 2.602 | 2.947 | 3.286 | 3.733 | 4.073 |
| 16 | 0.258 | 0.690 | 1.337 | 1.746 | 2.120 | 2.583 | 2.921 | 3.252 | 3.686 | 4.015 |
| 17 | 0.257 | 0.689 | 1.333 | 1.740 | 2.110 | 2.567 | 2.898 | 3.222 | 3.646 | 3.965 |
| 18 | 0.257 | 0.688 | 1.330 | 1.734 | 2.101 | 2.552 | 2.878 | 3.197 | 3.610 | 3.922 |
| 19 | 0.257 | 0.688 | 1.328 | 1.729 | 2.093 | 2.539 | 2.861 | 3.174 | 3.579 | 3.883 |
| 20 | 0.257 | 0.687 | 1.325 | 1.725 | 2.086 | 2.528 | 2.845 | 3.153 | 3.552 | 3.850 |
| 21 | 0.257 | 0.686 | 1.323 | 1.721 | 2.080 | 2.518 | 2.831 | 3.135 | 3.527 | 3.819 |
| 22 | 0.256 | 0.686 | 1.321 | 1.717 | 2.074 | 2.508 | 2.819 | 3.119 | 3.505 | 3.792 |
| 23 | 0.256 | 0.685 | 1.319 | 1.714 | 2.069 | 2.500 | 2.807 | 3.104 | 3.485 | 3.767 |
| 24 | 0.256 | 0.685 | 1.318 | 1.711 | 2.064 | 2.492 | 2.797 | 3.091 | 3.467 | 3.745 |
| 25 | 0.256 | 0.684 | 1.316 | 1.708 | 2.060 | 2.485 | 2.787 | 3.078 | 3.450 | 3.725 |
| 26 | 0.256 | 0.684 | 1.315 | 1.706 | 2.056 | 2.479 | 2.779 | 3.067 | 3.435 | 3.707 |
| 27 | 0.256 | 0.684 | 1.314 | 1.703 | 2.052 | 2.473 | 2.771 | 3.057 | 3.421 | 3.690 |
| 28 | 0.256 | 0.683 | 1.313 | 1.701 | 2.048 | 2.467 | 2.763 | 3.047 | 3.408 | 3.674 |
| 29 | 0.256 | 0.683 | 1.311 | 1.699 | 2.045 | 2.462 | 2.756 | 3.038 | 3.396 | 3.659 |
| 30 | 0.256 | 0.683 | 1.310 | 1.697 | 2.042 | 2.457 | 2.750 | 3.030 | 3.385 | 3.646 |
| 40 | 0.255 | 0.681 | 1.303 | 1.684 | 2.021 | 2.423 | 2.704 | 2.971 | 3.307 | 3.551 |
| 60 | 0.254 | 0.679 | 1.296 | 1.671 | 2.000 | 2.390 | 2.660 | 2.915 | 3.232 | 3.460 |
| 120 | 0.254 | 0.677 | 1.289 | 1.658 | 1.980 | 2.358 | 2.617 | 2.860 | 3.160 | 3.373 |
| ∞ | 0.253 | 0.674 | 1.282 | 1.645 | 1.960 | 2.326 | 2.576 | 2.807 | 3.090 | 3.291 |

The above t-table has a left-end column giving degrees of freedom (≈ sample sizes), and two top rows with p-values (areas under the curve = p – values), one-tail meaning that only one end of the curve, two-tail meaning that both ends are assessed simultaneously. The t-table is, furthermore, full of t-values, that, with $\infty$ degrees of freedom, are equal to z-values. The z-values and t-values are to be understood as mean results of studies, but not expressed in mmol/l, kilograms, or proportions of responders, but in so-called SEM-units (Standard error of the mean units), that are obtained by dividing your mean result by its own standard error. For continuous outcome data, with many degrees of freedom (large samples) the curve will be a little bit narrower, and more in agreement with nature. For binary outcome data, nature has determined that the curves will always be as narrow as can be, according to the row at the bottom.

# 6 Conclusion

Traditionally, $2 \times 2$ cross-tabs consist of two unpaired groups of patients. Sometimes, however, a single group is assessed twice, and, then, we will obtain a cross-tab slightly different from the traditional unpaired cross-tabs. McNemar's tests are helpful for analyzing these kinds of data. Either chi-square or z statistic can be used for testing. McNemar's tests are appropriate for the analysis of paired binary outcome data, like the data from diagnostic or therapeutic crossover studies.

# 7 Note

More background, theoretical and mathematical information of McNemar's tests is given in Statistics applied to clinical studies 5th edition, Chap. 3, Springer Heidelberg Germany, 2012, from the same authors.

# Chapter 49
# McNemar's Odds Ratios

## 1 General Purpose

The Chap. 44 has reviewed odds ratios (ORs) for analyzing cross-tabs of two unpaired groups of patients. Sometimes a single group is assessed twice, and, then, we will obtain a cross-tab slightly different from the traditional unpaired cross-tabs. McNemar's test must be applied for analyzing these kind of data (Chap. 48). Just like the odds ratios obtained from unpaired cross-tabs, odds ratios can be obtained from paired cross-tabs. It gives an estimate, in a crossover study of two treatments or two diagnostic tests, of, how much better one is than the other.

## 2 Schematic Overview of Type of Data File

| Outcome-1 binary | outcome-2 binary |
|---|---|
| . | . |
| . | . |
| . | . |
| . | . |
| . | . |
| . | . |
| . | . |
| . | . |
| . | . |
| | |
| | |
| | |
| | |



T.J. Cleophas, A.H. Zwinderman, *Clinical Data Analysis on a Pocket Calculator*,
DOI 10.1007/978-3-319-27104-0_49

## 3 Primary Scientific Question

In a crossover study two diagnostic devices are assessed. We wish to know not only, whether one performs better than the other, but also, how much better.

## 4 Data Example

Just like with the usual unpaired cross-tabs (Chap. 44), odds ratios can be calculated from crossover data with a single group of patients tested twice instead of two groups tested once. So far, we assessed two groups, and one treatment. Now, two antihypertensive treatments will be assessed in a single group of patients. OR = odds ratio, SE = standard error, and ln = natural logarithm.

normotension with drug 1

| | | yes | no |
|---|---|---|---|
| normotension | yes | (a) 65 | (b) 28 |
| with drug 2 | no | (c) 12 | (d) 34 |

Here the OR = b/c, and the SE is not $\sqrt{\left(\frac{1}{a}+\frac{1}{b}+\frac{1}{c}+\frac{1}{d}\right)}$, but rather $\sqrt{\left(\frac{1}{b}+\frac{1}{c}\right)}$.

OR = 28/12
= 2.33

This would mean that one treatment is about 2.33 times better than the other.

lnOR = ln 2.33
= 0.847

SE $= \sqrt{\left(\frac{1}{b}+\frac{1}{c}\right)} = 0.345$

lnOR ± 2 SE = 0.847 ± 0.690
= between 0.157 and 1.537,

Turn the ln numbers into real numbers by the anti-ln button (the invert button, on many calculators, called the 2ndF button) of your pocket calculator.

= between 1.16 and 4.65
= significantly different from 1.0.

A p-value can be calculated using the z-test with the help of the t-table.

z = lnOR/SEM
= 0.847:0.345
= 2.455.

| df | One-Tail = .4 | .25 | .1 | .05 | .025 | .01 | .005 | .0025 | .001 | .0005 |
|---|---|---|---|---|---|---|---|---|---|---|
| | Two-Tail = .8 | .5 | .2 | .1 | .05 | .02 | .01 | .005 | .002 | .001 |
| 1 | 0.325 | 1.000 | 3.078 | 6.314 | 12.706 | 31.821 | 63.657 | 127.32 | 318.31 | 636.62 |
| 2 | 0.289 | 0.816 | 1.886 | 2.920 | 4.303 | 6.965 | 9.925 | 14.089 | 22.327 | 31.598 |
| 3 | 0.277 | 0.765 | 1.638 | 2.353 | 3.182 | 4.541 | 5.841 | 7.453 | 10.214 | 12.924 |
| 4 | 0.271 | 0.741 | 1.533 | 2.132 | 2.776 | 3.747 | 4.604 | 5.598 | 7.173 | 8.610 |
| 5 | 0.267 | 0.727 | 1.476 | 2.015 | 2.571 | 3.365 | 4.032 | 4.773 | 5.893 | 6.869 |
| 6 | 0.265 | 0.718 | 1.440 | 1.943 | 2.447 | 3.143 | 3.707 | 4.317 | 5.208 | 5.959 |
| 7 | 0.263 | 0.711 | 1.415 | 1.895 | 2.365 | 2.998 | 3.499 | 4.029 | 4.785 | 5.408 |
| 8 | 0.262 | 0.706 | 1.397 | 1.860 | 2.306 | 2.896 | 3.355 | 3.833 | 4.501 | 5.041 |
| 9 | 0.261 | 0.703 | 1.383 | 1.833 | 2.262 | 2.821 | 3.250 | 3.690 | 4.297 | 4.781 |
| 10 | 0.260 | 0.700 | 1.372 | 1.812 | 2.228 | 2.764 | 3.169 | 3.581 | 4.144 | 4.587 |
| 11 | 0.260 | 0.697 | 1.363 | 1.796 | 2.201 | 2.718 | 3.106 | 3.497 | 4.025 | 4.437 |
| 12 | 0.259 | 0.695 | 1.356 | 1.782 | 2.179 | 2.681 | 3.055 | 3.428 | 3.930 | 4.318 |
| 13 | 0.259 | 0.694 | 1.350 | 1.771 | 2.160 | 2.650 | 3.012 | 3.372 | 3.852 | 4.221 |
| 14 | 0.258 | 0.692 | 1.345 | 1.761 | 2.145 | 2.624 | 2.977 | 3.326 | 3.787 | 4.140 |
| 15 | 0.258 | 0.691 | 1.341 | 1.753 | 2.131 | 2.602 | 2.947 | 3.286 | 3.733 | 4.073 |
| 16 | 0.258 | 0.690 | 1.337 | 1.746 | 2.120 | 2.583 | 2.921 | 3.252 | 3.686 | 4.015 |
| 17 | 0.257 | 0.689 | 1.333 | 1.740 | 2.110 | 2.567 | 2.898 | 3.222 | 3.646 | 3.965 |
| 18 | 0.257 | 0.688 | 1.330 | 1.734 | 2.101 | 2.552 | 2.878 | 3.197 | 3.610 | 3.922 |
| 19 | 0.257 | 0.688 | 1.328 | 1.729 | 2.093 | 2.539 | 2.861 | 3.174 | 3.579 | 3.883 |
| 20 | 0.257 | 0.687 | 1.325 | 1.725 | 2.086 | 2.528 | 2.845 | 3.153 | 3.552 | 3.850 |
| 21 | 0.257 | 0.686 | 1.323 | 1.721 | 2.080 | 2.518 | 2.831 | 3.135 | 3.527 | 3.819 |
| 22 | 0.256 | 0.686 | 1.321 | 1.717 | 2.074 | 2.508 | 2.819 | 3.119 | 3.505 | 3.792 |
| 23 | 0.256 | 0.685 | 1.319 | 1.714 | 2.069 | 2.500 | 2.807 | 3.104 | 3.485 | 3.767 |
| 24 | 0.256 | 0.685 | 1.318 | 1.711 | 2.064 | 2.492 | 2.797 | 3.091 | 3.467 | 3.745 |
| 25 | 0.256 | 0.684 | 1.316 | 1.708 | 2.060 | 2.485 | 2.787 | 3.078 | 3.450 | 3.725 |
| 26 | 0.256 | 0.684 | 1.315 | 1.706 | 2.056 | 2.479 | 2.779 | 3.067 | 3.435 | 3.707 |
| 27 | 0.256 | 0.684 | 1.314 | 1.703 | 2.052 | 2.473 | 2.771 | 3.057 | 3.421 | 3.690 |
| 28 | 0.256 | 0.683 | 1.313 | 1.701 | 2.048 | 2.467 | 2.763 | 3.047 | 3.408 | 3.674 |
| 29 | 0.256 | 0.683 | 1.311 | 1.699 | 2.045 | 2.462 | 2.756 | 3.038 | 3.396 | 3.659 |
| 30 | 0.256 | 0.683 | 1.310 | 1.697 | 2.042 | 2.457 | 2.750 | 3.030 | 3.385 | 3.646 |
| 40 | 0.255 | 0.681 | 1.303 | 1.684 | 2.021 | 2.423 | 2.704 | 2.971 | 3.307 | 3.551 |
| 60 | 0.254 | 0.679 | 1.296 | 1.671 | 2.000 | 2.390 | 2.660 | 2.915 | 3.232 | 3.460 |
| 120 | 0.254 | 0.677 | 1.289 | 1.658 | 1.980 | 2.358 | 2.617 | 2.860 | 3.160 | 3.373 |
| ∞ | 0.253 | 0.674 | 1.282 | 1.645 | 1.960 | 2.326 | 2.576 | 2.807 | 3.090 | 3.291 |

The t-table has a left-end column giving degrees of freedom (≈ sample sizes), and two top rows with p-values (areas under the curve = p – values), one-tail meaning that only one end of the curve, two-tail meaning that both ends are assessed simultaneously. The t-table is, furthermore, full of t-values, that, with ∞ degrees of freedom, are equal to z-values. The z-values and t-values are to be understood as mean results of studies, but not expressed in mmol/l, kilograms, or proportions of responders, but in so-called SEM-units (Standard error of the mean units), that are obtained by dividing your mean result by its own standard error. For continuous outcome data, with many degrees of freedom (large samples) the curve will be a little bit narrower, and more in agreement with nature. For binary outcome data, nature has determined that the curves will always be as narrow as can be, according to the row at the bottom.

The bottom row of the above t-table shows that this z-value is 2.455, and, thus, larger than 2.326. This means that the corresponding p-value of <0.02. The two drugs, thus, produce significantly different results at $p < 0.02$.

We may, additionally, conclude that one drug is about 2.33 times better than the other.

## 5 Conclusion

Odds ratios (ORs) are for analyzing cross-tabs of two unpaired groups of patients.

McNemar's odds ratios are for analyzing unpaired cross-tabs obtained from crossover data with binary outcome values. Just like the traditional odds ratios they give an estimate of how much better one treatment or one diagnostic device is than the other.

## 6 Note

More background, theoretical and mathematical information of McNemar's odds ratio testing is given in Statistics applied to clinical studies 5th edition, Chap. 3, Springer Heidelberg Germany, 2012, from the same authors.

# Chapter 50
# Power Equations

## 1 General Purpose

Power can be described as statistical conclusive force. It can be defined as the chance of finding a difference where there is one. Other chances are the chance of finding no difference where there is one (type II error) and the chance of finding a difference where there is none (type I) error. The result of a study with binary outcome data is often expressed in the form of the proportion of the responders or positive tests out of all patients/all tests, and its standard deviation (SD) or standard error (SE). With the proportion getting larger and the standard error getting smaller, the study obtains increasing power. This chapter is to show how to compute a study's statistical power from its main results.

## 2 Schematic Overview of Type of Data File

| Outcome (binary or yes-no) |
|---|
| . |
| . |
| . |
| . |
| . |
| . |
| . |
| . |
| . |
| |
| |
| |



T.J. Cleophas, A.H. Zwinderman, *Clinical Data Analysis on a Pocket Calculator*,
DOI 10.1007/978-3-319-27104-0_50

## 3 Primary Scientific Question

What is the power of a study with its proportion of responders and its standard error given.

## 4 Data Example

Only 4 out of 10 patients were responders to antihypertensive treatment. Is this result statistically significant. In other words is four responders significantly more than 0 responders?

The proportion responders = 4/10 = 0.4

The standard error of this proportion (SE) = √ (p (1−p)/n)
with p = proportion and n = sample size
= √ (0.4 × 0.6/10)
= √ (0.24/10)
= 0.1549

The t-test is used for test statistic.

t-value = proportion/SE
= 0.4/0.1549
= 2.582

This t-value, often called z-value with binary outcomes, is larger than 1.960 and, this means, that we can conclude that 4 out of 10 is significantly more than 0 out of 10 at $p < 0.05$ (see underneath t-table).

What is the power of this test? The equation below has to be applied.

$$\text{Power} = 1 - \text{prob}\left(z < t - t^1\right)$$

| | |
|---|---|
| t | = the t-value of your results = 2.582 |
| $t^1$ | = the t – value, that matches a p-value of 0.05 = 1.960; |
| t | = 2.582; $t^1$ = 1.960; $t - t^1$ = 0.622; this value is close to 0.674; |
| prob ($z < t - t^1$) | = beta = type II error = close to 0.25, maybe 0.30 |
| 1-beta = power | = close to 0.70 = close to 70 %. |

A power of 70 % is not very much. We have a chance of a type II error of 30 %. It, actually means, that, next time you perform the study, you will have about 30 % chance of an unsignificant result.

Explanation of the above calculation is given in the next few lines. The t-table on the next page is needed for the purpose. For binary outcome data only the bottom row is needed. The t-values are often, simply, called z-values here. The $t - t^1$ value of 0.622 is close to 0.674. Look right up at the top rows for finding beta (type II error = the chance of finding no difference where there is one). We have two top rows here, one for one-tail testing one for two-tail testing. Power is always tested one-tail. Null-hypothesis testing is mostly tested two-tail. In the t-table, we are a bit left from 0.25 (25 %), maybe 0.30 (30 %). This would be a pretty adequate estimate of the type II error, otherwise called beta of this small study. The power equals (100 % – beta) = close to 100 – 30 % = 70 %.

| df | One-Tail = .4<br>Two-Tail = .8 | .25<br>.5 | .1<br>.2 | .05<br>.1 | .025<br>.05 | .01<br>.02 | .005<br>.01 | .0025<br>.005 | .001<br>.002 | .0005<br>.001 |
|---|---|---|---|---|---|---|---|---|---|---|
| 1 | 0.325 | 1.000 | 3.078 | 6.314 | 12.706 | 31.821 | 63.657 | 127.32 | 318.31 | 636.62 |
| 2 | 0.289 | 0.816 | 1.886 | 2.920 | 4.303 | 6.965 | 9.925 | 14.089 | 22.327 | 31.598 |
| 3 | 0.277 | 0.765 | 1.638 | 2.353 | 3.182 | 4.541 | 5.841 | 7.453 | 10.214 | 12.924 |
| 4 | 0.271 | 0.741 | 1.533 | 2.132 | 2.776 | 3.747 | 4.604 | 5.598 | 7.173 | 8.610 |
| 5 | 0.267 | 0.727 | 1.476 | 2.015 | 2.571 | 3.365 | 4.032 | 4.773 | 5.893 | 6.869 |
| 6 | 0.265 | 0.718 | 1.440 | 1.943 | 2.447 | 3.143 | 3.707 | 4.317 | 5.208 | 5.959 |
| 7 | 0.263 | 0.711 | 1.415 | 1.895 | 2.365 | 2.998 | 3.499 | 4.029 | 4.785 | 5.408 |
| 8 | 0.262 | 0.706 | 1.397 | 1.860 | 2.306 | 2.896 | 3.355 | 3.833 | 4.501 | 5.041 |
| 9 | 0.261 | 0.703 | 1.383 | 1.833 | 2.262 | 2.821 | 3.250 | 3.690 | 4.297 | 4.781 |
| 10 | 0.260 | 0.700 | 1.372 | 1.812 | 2.228 | 2.764 | 3.169 | 3.581 | 4.144 | 4.587 |
| 11 | 0.260 | 0.697 | 1.363 | 1.796 | 2.201 | 2.718 | 3.106 | 3.497 | 4.025 | 4.437 |
| 12 | 0.259 | 0.695 | 1.356 | 1.782 | 2.179 | 2.681 | 3.055 | 3.428 | 3.930 | 4.318 |
| 13 | 0.259 | 0.694 | 1.350 | 1.771 | 2.160 | 2.650 | 3.012 | 3.372 | 3.852 | 4.221 |
| 14 | 0.258 | 0.692 | 1.345 | 1.761 | 2.145 | 2.624 | 2.977 | 3.326 | 3.787 | 4.140 |
| 15 | 0.258 | 0.691 | 1.341 | 1.753 | 2.131 | 2.602 | 2.947 | 3.286 | 3.733 | 4.073 |
| 16 | 0.258 | 0.690 | 1.337 | 1.746 | 2.120 | 2.583 | 2.921 | 3.252 | 3.686 | 4.015 |
| 17 | 0.257 | 0.689 | 1.333 | 1.740 | 2.110 | 2.567 | 2.898 | 3.222 | 3.646 | 3.965 |
| 18 | 0.257 | 0.688 | 1.330 | 1.734 | 2.101 | 2.552 | 2.878 | 3.197 | 3.610 | 3.922 |
| 19 | 0.257 | 0.688 | 1.328 | 1.729 | 2.093 | 2.539 | 2.861 | 3.174 | 3.579 | 3.883 |
| 20 | 0.257 | 0.687 | 1.325 | 1.725 | 2.086 | 2.528 | 2.845 | 3.153 | 3.552 | 3.850 |
| 21 | 0.257 | 0.686 | 1.323 | 1.721 | 2.080 | 2.518 | 2.831 | 3.135 | 3.527 | 3.819 |
| 22 | 0.256 | 0.686 | 1.321 | 1.717 | 2.074 | 2.508 | 2.819 | 3.119 | 3.505 | 3.792 |
| 23 | 0.256 | 0.685 | 1.319 | 1.714 | 2.069 | 2.500 | 2.807 | 3.104 | 3.485 | 3.767 |
| 24 | 0.256 | 0.685 | 1.318 | 1.711 | 2.064 | 2.492 | 2.797 | 3.091 | 3.467 | 3.745 |
| 25 | 0.256 | 0.684 | 1.316 | 1.708 | 2.060 | 2.485 | 2.787 | 3.078 | 3.450 | 3.725 |
| 26 | 0.256 | 0.684 | 1.315 | 1.706 | 2.056 | 2.479 | 2.779 | 3.067 | 3.435 | 3.707 |
| 27 | 0.256 | 0.684 | 1.314 | 1.703 | 2.052 | 2.473 | 2.771 | 3.057 | 3.421 | 3.690 |
| 28 | 0.256 | 0.683 | 1.313 | 1.701 | 2.048 | 2.467 | 2.763 | 3.047 | 3.408 | 3.674 |
| 29 | 0.256 | 0.683 | 1.311 | 1.699 | 2.045 | 2.462 | 2.756 | 3.038 | 3.396 | 3.659 |
| 30 | 0.256 | 0.683 | 1.310 | 1.697 | 2.042 | 2.457 | 2.750 | 3.030 | 3.385 | 3.646 |
| 40 | 0.255 | 0.681 | 1.303 | 1.684 | 2.021 | 2.423 | 2.704 | 2.971 | 3.307 | 3.551 |
| 60 | 0.254 | 0.679 | 1.296 | 1.671 | 2.000 | 2.390 | 2.660 | 2.915 | 3.232 | 3.460 |
| 120 | 0.254 | 0.677 | 1.289 | 1.658 | 1.980 | 2.358 | 2.617 | 2.860 | 3.160 | 3.373 |
| ∞ | 0.253 | 0.674 | 1.282 | 1.645 | 1.960 | 2.326 | 2.576 | 2.807 | 3.090 | 3.291 |

The t-table has a left-end column giving degrees of freedom (≈ sample sizes), and two top rows with p-values (areas under the curve = p – values), one-tail meaning that only one end of the curve, two-tail meaning that both ends are assessed simultaneously. The t-table is, furthermore, full of t-values, that, with $\infty$ degrees of freedom, are equal to z-values. The z-values and t-values are to be understood as mean results of studies, but not expressed in mmol/l, kilograms, or proportions of responders, but in so-called SEM-units (Standard error of the mean units), that are obtained by dividing your mean result by its own standard error. For continuous outcome data, with many degrees of freedom (large samples) the curve will be a little bit narrower, and more in agreement with nature. For binary outcome data, nature has determined that the curves will always be as narrow as can be, according to the row at the bottom.

## 5 Conclusion

Power can be defined as the chance of finding a difference, where there is one. It is equal to 1 minus the type II error ($= 1 - \beta$). A study result with binary outcome data is often expressed in the form of proportion of responders and its standard deviation (SD) or standard error (SE). With this proportion getting larger and the standard error getting smaller, the study will obtain increasing power. This chapter shows, how to compute a study's statistical power from its proportion responders and standard error. We recommend to read Chap. 11 for a better understanding of the reasoning of the procedure in the current chapter.

## 6 Note

More background, theoretical and mathematical information of power assessments is given in Statistics applied to clinical studies 5th edition, Chap. 6, Springer Heidelberg Germany, 2012, from the same authors.

# Chapter 51
# Sample Size Calculations

## 1 General Purpose

Just like with continuous outcome data (Chap. 12), with binary outcome data an essential part of the study protocol is the assessment of the question, how many subject need to be studied in order to answer the studies' objectives. This chapter provides equations that can be used for the purpose.

## 2 Schematic Overview of Type of Data File

| Outcome (yes, no) |
|---|
| . |
| . |
| . |
| . |
| . |
| . |
| . |
| . |
| . |
| |
| |
| |



T.J. Cleophas, A.H. Zwinderman, *Clinical Data Analysis on a Pocket Calculator*,
DOI 10.1007/978-3-319-27104-0_51

## 3 Primary Scientific Question

What sample size do we need in order to produce a study with a statistically significant result?

## 4 Data Example, Binary Data, Power 80 %

What is the required sample size of a study in which you expect an event in 10 % of the patients and wish to have a power of 80 %.

10 % events means a proportion of events of 0.1.

The standard deviation (SD) of this proportion is defined by the equation

$\sqrt{}$ [proportion × (1 – proportion)] =

$\sqrt{}$ (0.1 × 0.9).

The suitable formula is given.

Required sample size $= \text{power index} \times SD^2/\text{proportie}^2$
$= 7.8 \times (0.1 \times 0.9)/0.1^2$
$= 7.8 \times 9 = 71.$

We conclude that with 10 % events you will need about 71 patients in order to obtain a significant number of events for a power of 80 % in your study.

## 5 Data Example, Binary Data, Power 80 %, Two Groups

What is the required sample size of a study of two groups in which you expect

A difference in events between the two groups of 10 %, and in which you wish to have a power of 80 %.

10 % difference in events means a difference in proportions of events of 0.10.

Let us assume that in Group one 10 % will have an event and in Group two 20 %. The standard deviations per group can be calculated.

For group 1: SD = $\sqrt{}$ [proportion × (1 – proportion)] $= \sqrt{(0.1 \times 0.9)} = 0.3$
For group 2: SD = $\sqrt{}$ [proportion × (1 – proportion)] $= \sqrt{(0.2 \times 0.8)} = 0.4$
The pooled standard deviation of both groups $= \sqrt{(SD_1{}^2 + SD_2{}^2)}$
$= \sqrt{(0.3^2 + 0.4^2)}$
$= \sqrt{0.25} = 0.5$

The adequate equation is underneath.

Required sample size $= \text{power index} \times (\text{pooled SD})^2/(\text{difference in proportions})^2$
$= 7.8 \times 0.5^2/0.1^2$

$$= 7.8 \times 25$$
$$= 195.$$

Obviously, with a difference of 10 % events between two groups we will need about 195 patients per group in order to demonstrate a significant difference with a power of 80 %.

## 6 Conclusion

The assessment of the sample size required to adequately answer a study's objectives is an essential part of the study protocol. This is true both for studies with continuous (Chap. 12), and those with binary outcome data (the current chapter). Equations are given for studies with binary outcome data.

## 7 Note

More background, theoretical and mathematical information of sample size assessments is given in Statistics applied to clinical studies 5th edition, Chap. 6, Springer Heidelberg Germany, 2012, from the same authors.

# Chapter 52
# Accuracy Assessments

## 1 General Purpose

Sensitivity and specificity are measures of diagnostic accuracy of qualitative diagnostic tests, and are obtained from data samples. Just like averages, they are estimates and come with certain amounts of uncertainty. The STARDS (Standards for Reporting Diagnostic Accuracy) working party recommends to include measures of uncertainty in any evaluation of a diagnostic test. This chapter assesses how accuracy measures for qualitative diagnostic tests can be tested for uncertainty.

## 2 Schematic Overview of Type of Data File

| Definitive diagnosis (yes-no) | Diagnostic test (yes-no) |
|---|---|
| . | . |
| . | . |
| . | . |
| . | . |
| . | . |
| . | . |
| . | . |
| . | . |
| . | . |
| | |
| | |
| | |



T.J. Cleophas, A.H. Zwinderman, *Clinical Data Analysis on a Pocket Calculator*,
DOI 10.1007/978-3-319-27104-0_52

## 3 Primary Scientific Question

How do we assess accuracy measures for qualitative diagnostic tests can be tested for uncertainty.

## 4 Estimating Uncertainty of Sensitivity and Specificity

For the calculation of the standard errors (SEs) of sensitivity, specificity and overall-validity we make use of the Gaussian curve assumption in the data.

| | | Definitive diagnosis (n) | |
|---|---|---|---|
| | | Yes | No |
| Result diagnostic test | Yes | a | b |
| | No | c | d |

Sensitivity = a/(a + c) = proportion true positives
Specificity = d/(b + d) = proportion true negatives
1−specificity = b/(b + d)
Proportion of patients with a definitive diagnosis = (a + c)/(a + b + c + d)
Overall validity = (a + d)/(a + b + c + d)

In order to make predictions from these estimates of validity their standard deviations / errors are required. The standard deviation / error (SD/ SE) of a proportion can be calculated.

| | |
|---|---|
| SD = | √ p(1 − p) where p = proportion. |
| SE = | √ [p(1 − p)/n] where n = sample size |

where p equals a/(a + c) for the sensitivity. Using the above equations de standard errors can be readily obtained.

| | |
|---|---|
| $SE_{sensitivity}$ = | $\sqrt{ac/(a+c)^3}$ |
| $SE_{specificity}$ = | $\sqrt{db/(d+b)^3}$ |
| $SE_{1-specificity}$ = | $\sqrt{db/(d+b)^3}$ |
| $SE_{proportion\ of\ patients\ with\ a\ definitive\ diagnosis}$ = | $\sqrt{(a+b)(c+d)/(a+b+c+d)^3}$ |

## 5 Example 1

Two hundred patients are evaluated the determine the sensitivity/specificity of B-type Natriuretic Peptide (BNP) for making a diagnosis of heart failure.

| | | Heart failure (n) | |
|---|---|---|---|
| | | Yes | No |
| Result diagnostic test | positive | 70 (a) | 35 (b) |
| | negative | 30 (c) | 65 (d) |

The sensitivity (a/(a + c)) and specificity (d/(b + d)) are calculated to be 0.70 and 0.65 respectively (70 and 65 %). In order for these estimates to be significantly larger than 50 % their 95 % confidence interval should not cross the 50 % boundary.

The standard errors are calculated using the above equations. For sensitivity the standard error is 0.0458, for specificity 0.0477. Under the assumption of Gaussian curve distributions in the data the 95 % confidence intervals of the sensitivity and specificity can be calculated according to:

95 % confidence interval of the sensitivity $= 0.70 \pm 1.96 \times 0.0458$
" " " specificity $= 0.65 \pm 1.96 \times 0.0477.$

This means that the 95 % confidence interval of the sensitivity is between 61 % and 79 %, for specificity it is between 56 % and 74 %. These results do not cross the 50 % boundary and fall, thus, entirely within the boundary of validity. The diagnostic test can be accepted as being valid.

# 6 Example 2

Dimer tests have been widely used as screening tests for lung embolias.

| | | Lung embolia (n) | |
|---|---|---|---|
| | | Yes | No |
| Dimer test | Positive | 2 (a) | 18 (b) |
| | Negative | 1 (c) | 182 (d) |

The sensitivity (a/(a + c)) and specificity (d/(b + d)) are calculated to be 0.666 and 0.911 respectively (67 and 91 %). In order for these estimates to be significantly larger than 50 % the 95 % confidence interval of them should again not cross the 50 % boundary.

The standard errors, as calculated according to the above equations, are for sensitivity 0.272, for specificity 0.040. Under the assumption of Gaussian curve distributions the 95 % confidence intervals of the sensitivity and specificity are calculated according to:

95 % confidence interval of the sensitivity $= 0.67 \pm 1.96 \times 0.272$
" " " specificity $= 0.91 \pm 1.96 \times 0.040.$

The 95 % confidence interval of the sensitivity is between 0.14 and 1.20 (14 and 120 %). The 95 % confidence interval of the specificity can be similarly calculated, and is between 0.87 and 0.95 (87 and 95 %). The interval for the sensitivity is very wide and does not at all fall within the boundaries of 0.5–1.0 (50–100 %). Validity of this test is, therefore, not really demonstrated. The appropriate conclusion of this evaluation should be: based on this evaluation the diagnostic cannot be accepted as being valid in spite of a sensitivity and specificity of respectively 67 and 91 %.

## 7 Conclusion

Sensitivity and specificity are measures of diagnostic accuracy of qualitative diagnostic tests. Just like averages they are estimates and come with certain amounts of uncertainty. This chapter assesses how accuracy measures for qualitative diagnostic tests can be tested for uncertainty.

## 8 Note

More background, theoretical and mathematical information of accuracy assessments of qualitative diagnostic tests is given in Statistics applied to clinical studies 5th edition, Chaps. 46 and 47, Springer Heidelberg Germany, 2012, from the same authors.

# Chapter 53
# Reliability Assessments

## 1 General Purpose

The reproducibility, otherwise called reliability, of continuous data can be estimated with duplicate standard deviations (Chap. 26). With binary data Cohen's kappas are used for the purpose. Reliability assessment of diagnostic procedures is an important part of the validity assessment of scientific research. The current chapter shows how it works.

## 2 Schematic Overview of Type of Data File

| Outcome first test | Outcome second test |
|---|---|
| . | . |
| . | . |
| . | . |
| . | . |
| . | . |
| . | . |
| . | . |
| . | . |
| . | . |
| | |
| | |
| | |
| | |
| | |



T.J. Cleophas, A.H. Zwinderman, *Clinical Data Analysis on a Pocket Calculator*,
DOI 10.1007/978-3-319-27104-0_53

## 3 Primary Scientific Question

How can binary outcome data from a diagnostic test be tested for reproducibility, otherwise called reliability.

## 4 Example

Positive (pos) or negative (neg) laboratory tests of 30 patients are assessed. All patients are tested a second time in order to estimate the level of reproducibility of the test.

| | | 1st time | | |
|---|---|---|---|---|
| | | pos | neg | |
| 2nd time | pos | 10 | 5 | 15 |
| | neg | 4 | 11 | 15 |
| | | 14 | 16 | 30 |

If the test is not reproducible at all, then we will find twice the same result in 50 % of the patients, and a different result the second time in the other 50 % of the patients.

Overall 30 tests have been carried out twice.

We observe 10 times 2 × positive and
11 times 2 × negative.

And, thus, twice the same is found in

21 patients which is considerable more than in half of the cases, which should have been 15 times.

Minimal indicates the number of duplicate observations if reproducibility were zero, maximal indicates the number of duplicate observations if the reproducibility were 100 %.

$$\text{Kappa} = \frac{\text{observed}-\text{minimal}}{\text{maximal}-\text{minimal}} = \frac{21-15}{30-15} = 0.4$$

A kappa-value of 0.0 means that reproducibility is very poor.

A kappa of 1.0 would have meant excellent reproducibility.

In our example we observed a kappa of 0.4, which means reproducibility is very moderate.

## 5 Conclusion

The reproducibility, otherwise called reliability, of continuous data can be estimated with duplicate standard deviations. With binary data Cohen's kappas are adequate for the purpose. Reliability assessment of diagnostic procedures is an important part of the validity assessment of scientific research.

## 6 Note

More background, theoretical and mathematical information of reliability assessments of binary diagnostic tests is given in Statistics applied to clinical studies 5th edition, Chap. 45, Springer Heidelberg Germany, 2012, from the same authors.

# Chapter 54
# Unmasking Fudged Data

## 1 General Purpose

Statistics is not good at detecting manipulated data. However, tests for randomness is possible. For example, the chi-square goodness of fit and the Kolmogorov Smirnov test are examples (Chap. 33). Data may, of course, be unrandom due to extreme inclusion criteria or inadequate data cleaning. But data fudging is another possibility. This chapter assesses the use of the final digits of the pattern of the numerical results of your study as a possible method for detecting unrandom or fudged data.

## 2 Schematic Overview of Type of Data File

| Final digits of relative risks |
|---|
| . |
| . |
| . |
| . |
| . |
| . |
| . |
| . |
| . |
| |
| |
| |



T.J. Cleophas, A.H. Zwinderman, *Clinical Data Analysis on a Pocket Calculator*,
DOI 10.1007/978-3-319-27104-0_54

## 3 Primary Scientific Question

How can the final digits of results of the study results show unrandomnes in your data.

## 4 Data Example

In a statin trial 96 risk ratios (RR's) were the main results of the study. Often 9 or 1 were observed as final digits: for example risk ratios like 0.99 or 0.89 or 1.01 or 1.011. The accuracy of these risk ratios were checked according to the underneath procedure.

| Final digit of RR | observed frequency | expected frequency | Σ(observed − expected)$^2$/expected |
|---|---|---|---|
| 0 | 24 | 9.6 | 21.6 |
| 1 | 39 | 9.6 | 90.0 |
| 2 | 3 | 9.6 | 4.5 |
| 3 | 0 | 9.6 | 9.6 |
| 4 | 0 | 9.6 | 9.6 |
| 5 | 0 | 9.6 | 9.6 |
| 6 | 0 | 9.6 | 9.6 |
| 7 | 1 | 9.6 | 7.7 |
| 8 | 2 | 9.6 | 6.0 |
| 9 | 27 | 9.6 | 31.5 |
| Total | 96 | 96.0 | 199.7 |

The above differences between observed and expected frequencies were tested with the multiple groups chi-square test (see also Chap. 42). With

$$\Sigma(\text{ observed} - \text{ expected})^2/\text{expected} = 199.7, \text{ and } 10 - 1 = 9 \text{ degrees of freedom,}$$

the difference between observed and expected was much larger than could happen by chance.

The underneath chi-square table has an upper row with areas under the curve (p-values), a left-end column with degrees of freedom (df), and a whole lot of chi-square values. It shows that, with $10-1=9$ degrees of freedom, our p-value will be $<0.001$, if the chi-square value is $>27.877$. Our chi-square value was 199.7.

Chi-squared distribution

| | Two-tailed *P*-value | | | |
|---|---|---|---|---|
| *df* | 0.10 | 0.05 | 0.01 | 0.001 |
| 1 | 2.706 | 3.841 | 6.635 | 10.827 |
| 2 | 4.605 | 5.991 | 9.210 | 13.815 |
| 3 | 6.251 | 7.851 | 11.345 | 16.266 |
| 4 | 7.779 | 9.488 | 13.277 | 18.466 |
| 5 | 9.236 | 11.070 | 15.086 | 20.515 |
| 6 | 10.645 | 12.592 | 16.812 | 22.457 |
| 7 | 12.017 | 14.067 | 18.475 | 24.321 |
| 8 | 13.362 | 15.507 | 20.090 | 26.124 |
| 9 | 14.684 | 16.919 | 21.666 | 27.877 |
| 10 | 15.987 | 18.307 | 23.209 | 29.588 |
| 11 | 17.275 | 19.675 | 24.725 | 31.264 |
| 12 | 18.549 | 21.026 | 26.217 | 32.909 |
| 13 | 19.812 | 22.362 | 27.688 | 34.527 |
| 14 | 21.064 | 23.685 | 29.141 | 36.124 |
| 15 | 22.307 | 24.996 | 30.578 | 37.698 |
| 16 | 23.542 | 26.296 | 32.000 | 39.252 |
| 17 | 24.769 | 27.587 | 33.409 | 40.791 |
| 18 | 25.989 | 28.869 | 34.805 | 42.312 |
| 19 | 27.204 | 30.144 | 36.191 | 43.819 |
| 20 | 28.412 | 31.410 | 37.566 | 45.314 |
| 21 | 29.615 | 32.671 | 38.932 | 46.796 |
| 22 | 30.813 | 33.924 | 40.289 | 48.268 |
| 23 | 32.007 | 35.172 | 41.638 | 49.728 |
| 24 | 33.196 | 36.415 | 42.980 | 51.179 |
| 25 | 34.382 | 37.652 | 44.314 | 52.619 |
| 26 | 35.536 | 38.885 | 45.642 | 54.051 |
| 27 | 36.741 | 40.113 | 46.963 | 55.475 |
| 28 | 37.916 | 41.337 | 48.278 | 56.892 |
| 29 | 39.087 | 42.557 | 49.588 | 58.301 |
| 30 | 40.256 | 43.773 | 50.892 | 59.702 |
| 40 | 51.805 | 55.758 | 63.691 | 73.403 |
| 50 | 63.167 | 67.505 | 76.154 | 86.660 |
| 60 | 74.397 | 79.082 | 88.379 | 99.608 |
| 70 | 85.527 | 90.531 | 100.43 | 112.32 |
| 80 | 96.578 | 101.88 | 112.33 | 124.84 |
| 90 | 107.57 | 113.15 | 124.12 | 137.21 |
| 100 | 118.50 | 124.34 | 135.81 | 149.45 |

And, so, the probability <0.001 that such a large chi-square value would occur by chance, and a chance finding can, thus, be rejected. We can conclude here, that the frequency distribution of final digits were not random. The validity of this trial is in jeopardy.

## 5 Conclusion

Statistics is not good at detecting manipulated data. But, the final digits of the pattern of the numerical results of your study as a possible method for detecting unrandom or fudged data.

## 6 Note

More background, theoretical and mathematical information of data unrandomness is given in Statistics applied to clinical studies 5th edition, Chaps. 42 and 43, Springer Heidelberg Germany, 2012, from the same authors.

# Chapter 55
# Markov modeling for Predicting Outside the Range of Observations

## 1 General Purpose

Regression models are only valid within the range of the x-values observed in the data. Markov modeling goes one step further, and aims at predicting outside the range of x-values. Like with Cox regression it assumes an exponential-pattern in the data which may be a strong assumption for complex human beings. This chapter gives examples.

## 2 Schematic Overview of Type of Data File

| Outcome binary | Predictor binary | Predictor binary | Predictor binary |
|---|---|---|---|
| . | . | . | |
| . | . | . | . |
| . | . | . | . |
| . | . | . | . |
| . | . | . | . |
| . | . | . | . |
| . | . | . | . |
| . | . | . | . |
| . | . | . | . |
| | | | |
| | | | |
| | | | |
| | | | |
| | | | |
| | | | |



T.J. Cleophas, A.H. Zwinderman, *Clinical Data Analysis on a Pocket Calculator*,
DOI 10.1007/978-3-319-27104-0_55

## 3 Primary Scientific Question

Can Markov modeling adequately be used for making predictions outside the range of observations.

## 4 Example 1

In patients with diabetes mellitus type II, sulfonureas are highly efficacious, but they will, eventually, induce beta-cell failure. Beta-cell failure is defined as a fasting plasma glucose >7.0 mmol/l. The question is, does the severity of diabetes and/or the potency of the sulfonurea-compound influence the induction of beta-cell failure? This was studied in 500 patients with diabetes type II.

at time 0 year 0/500 patients had beta-cell failure
at time 1 year 50/500 patients (= 10 % ) had beta-cell failure.

As after 1 year 90 % had no beta-cell failure, it is appropriate according to the Markow model to extrapolate:

after 2 years 90 % × 90 % = 81 % no beta-cell failure
after 3 years 90 % × 90 % × 90 % = 73 % no beta-cell failure
after 6.58 years = 50 % no beta-cell failure.

The calculation uses logarithmic transformation. We will use the example from the Chap. 43. In patients with diabetes mellitus (* = sign of multiplication):

After 1 year 10 % has beta-cell failure, and 90 % has not.
2 90 * 90 = 81 % has not.
3 90 * 90 * 90 = 73 % has not.

When will 50 % have beta-cell failure?
$0.9^x = 0.5$
$x \log 0.9 = \log 0.5$
$x = \log 0.5/\log 0.9 = 6.5788$ years.
For computation command in the Scientific Calculator the following.

Press 0.5....press log....press ÷button....press (....press 0.9....press log....press)....
press = .
The outcome displays 6.578813...

It will take around 6.6 years, before 50 % has beta-cell failure.

## 5 Example 2

A second question was, does the severity of diabetes mellitus type II influence induction of beta-cell failure. A cut-off level for severity often applied is a fasting plasma glucose > 10 mmol/l. According to the Markov modeling approach the question can be answered as follows:

250 patients had fasting plasma glucose < 10 mmol/l at diagnosis (Group-1)
250 patients had fasting plasma glucose > 10 mmol/l at diagnosis (Group-2)

If after 1 year sulfonureas (su) treatment, 10/250 of the patients from Group −1 had b-cell failure, and 40/250 of the patients from Group-2, which is significantly different with an odds ratio of 0.22 ( $p < 0.01$, see Chap. 44 if you wish to check the statistical test of this odds ratio).

$(240/250)^x = 0.5$
$x = \log 0.5/\log (24/25)$
$x = 16.9797..$

$(210/250)^x = 0.5$
$x = \log 0.5/\log (21/25)$
$x = 3.9755..$

The commands can be given as shown in the above section.
We conclude that:

in Group-1 it takes 17 years before 50 % of the patients develop beta-cell failure,
in Group-2 it takes 4 years before 50 % of the patients develop beta-cell failure.

## 6 Example 3

The next question is, does potency of su-compound influence induction of b-cell failure?

250 patients started on amaryl (potent sulfonurea) at diagnosis (Group-A)
250 patients started on artosin (non-potent sulfonurea) at diagnosis (Group-B)

If after 1 year 25/250 of Group-A had beta-cell failure, and 25/250 of the group-B, it is appropriate according to the Markov model to conclude that a non-potent does not prevent beta-cell failure.

## 7 Conclusion

Regression models are only valid within the range of the x-values observed in the data. Markov modeling goes one step further, and aims at predicting outside the range of x-values. Like with Cox regression it assumes an exponential-pattern. Markov modeling, although its use is very common in long-term observational studies remains highly speculative, because nature does not routinely follow mathematical models.

## 8 Note

More background, theoretical and mathematical information of is given in Statistics applied to clinical studies 5th edition, Chap. 17, Springer Heidelberg Germany, 2012, and Machine learning in medicine a complete overview, Chap. 55, Springer Heidelberg Germany, 2015, both from the same authors.

# Chapter 56
# Binary Partitioning for CART (Classification and Regression Tree) Methods

## 1 General Purpose

Binary partitioning is used to determine the best fit decision cut-off levels for a data set with false positive and false negative patients. It serves a purpose similar to that of the receiver operating characteristic (ROC) curve method (Cleophas and Zwinderman, SPSS for starters and second levelers, Chap. 46, Springer, New York, 2015), but, unlike ROC curves, it is adjusted for the magnitude of the samples, and therefore more precise. A hypothesized example is given in the underneath graph.

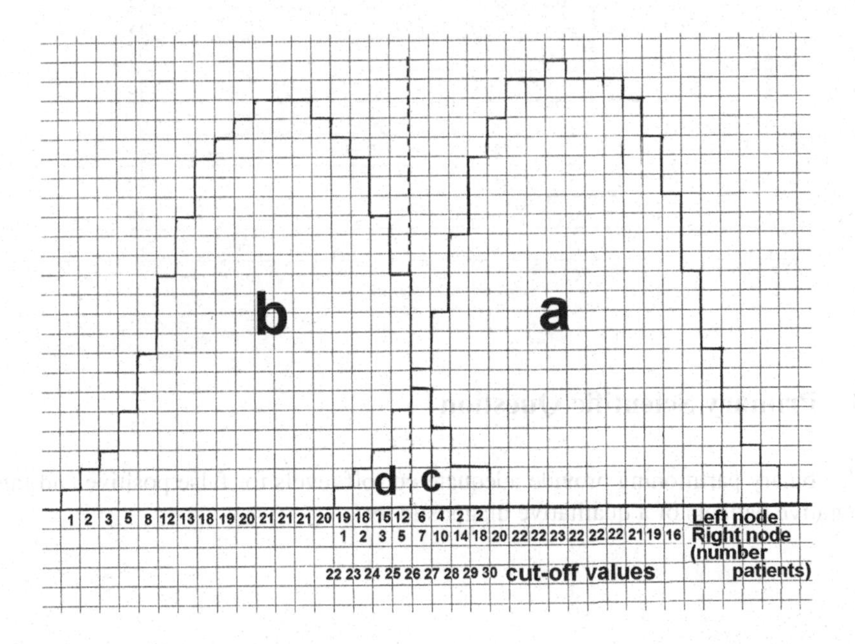




T.J. Cleophas, A.H. Zwinderman, *Clinical Data Analysis on a Pocket Calculator*,
DOI 10.1007/978-3-319-27104-0_56

The above histogram gives a patients' sample assessed for peripheral vascular disease; "a" summarizes the patients with a positive test and the presence of disease, "b" the patients with a negative test and the absence of disease, "c" and "d" are the false positive and false negative patients respectively. With binary partitioning, otherwise called the entropy method or CART (classification and regression tree) method, the entire sample of patients is called the parent node, which can, subsequently, be repeatedly split, partitioned if you will, into binary internal nodes. Mostly, internal nodes contain false positive or false negative patients, and are, thus, somewhat impure. The magnitude of their impurity is assessed. This chapter shows how it works.

## 2 Schematic Overview of Type of Data File

| Outcome binary | predictor binary |
|---|---|
| . | . |
| . | . |
| . | . |
| . | . |
| . | . |
| . | . |
| . | . |
| . | . |
| . | . |
| | |
| | |
| | |
| | |
| | |

## 3 Primary Scientific Question

Can binary partitioning provide adequate cut-off levels for false positive and false negative patients of a qualitative diagnostic test.

# 4 Binary Partitioning, the Analysis

With binary partitioning, otherwise called the entropy method or CART (classification and regression tree) method, the entire sample of patients (above graph) is called the parent node, which can, subsequently, be repeatedly split into binary internal nodes. Mostly, internal nodes contain false positive or negative patients, and are, thus, impure. The magnitude of their impurity is assessed by the log likelihood method (see also Chap. 46). Impurity equals the maximum log likelihood of the y-axis-variable by assuming that the x-axis-variable follows a Gaussian (i.e., binomial) distribution and is expressed in units, sometimes called bits (a short-cut for "binary digits"). All this sounds rather complex, but it works smoothly.

The x – axis variable for the right node $= x_r = a / (a+b)$,
for the left node $= x_l = d / (d+c)$.

If the impurity equals 1.0 bits, then it is maximal, if it equals 0.0, then it is minimal.

Impurity node either right or left $= -x \ln x - (1-x) \ln (1-x)$,
where ln means natural logarithm.

The impurities of the right and left node are calculated separately. Then, a weighted overall impurity of each cut-off level situation is calculated according to (* = sign of multiplication):

Weighted impurity cut-off =
[(a + b)/(a + b + c + d) * impurity-right-node] +
[(d + c)/(a + b + c + d) * impurity-left-node].

Underneath, an overview is given of the calculated impurities at the different cut-off levels. The cut-off percentage of 27 gives the smallest weighted impurity, and is, thus, the best fit predictor for the presence of peripheral vascular disease.

| Cut-off | impurity right node | impurity left node | impurity weighted |
|---|---|---|---|
| 22 % | 0.5137 | 0.0000 | 0.3180 |
| 23 % | 0.4392 | 0.0559 | 0.3063 |
| 24 % | 0.4053 | 0.0982 | 0.2766 |
| 25 % | 0.3468 | 0.1352 | 0.2711 |
| 26 % | 0.1988 | 0.1688 | 0.1897 |
| **27 %** | **0.1352** | **0.2268** | **0.1830** |
| 28 % | 0.0559 | 0.3025 | 0.1850 |
| 29 % | 0.0559 | 0.3850 | 0.2375 |
| 30 % | 0.0000 | 0.4690 | 0.2748 |

From the above calculation it can be concluded that a cut-off of 27 % is the best fit decision cut-off level with fewest false positive and fewest false negative patients. The result was slightly different from that of the ROC curve analysis, which produced a cut-off level of 26 %.

## 5 Conclusion

Binary partitioning is used to determine the best fit decision cut-off levels for a data set with false positive and false negative patients. It serves a purpose similar to that of the receiver operating characteristic (ROC) curve method., but, unlike ROC curves, it is adjusted for the magnitude of the samples, and therefore more precise. This chapter shows that the method works very well.

## 6 Note

More background, theoretical and mathematical information of binary partitioning is given in Statistics applied to clinical studies 5th edition, Chap. 53, Springer Heidelberg Germany, 2012, from the same authors.

# Chapter 57
# Meta-analysis of Binary Data

## 1 General Purpose

Meta-analyses can be defined as systematic reviews with pooled data. Because the separate studies in a meta-analysis have different sample sizes for the overall results, a weighted average has to be calculated. Heterogeneity in a meta-analysis means, that the differences in the results between the studies are larger than could happen by chance. The calculation of the overall result and the test for heterogeneity is demonstrated in the current chapter.

## 2 Schematic Overview of Type of Data File

| Outcome | predictor | patient characteristic.... |
|---|---|---|
| . | . | . |
| . | . | . |
| . | . | . |
| . | . | . |
| . | . | . |
| . | . | . |
| . | . | . |
| . | . | . |
| . | . | . |
| | | |
| | | |
| | | |
| | | |



T.J. Cleophas, A.H. Zwinderman, *Clinical Data Analysis on a Pocket Calculator*,
DOI 10.1007/978-3-319-27104-0_57

## 3 Primary Scientific Question

How do we assess pooled results of multiple studies, how do we assess heterogeneity between the studies.

## 4 Data Example, Pooling

The underneath data show the results of 7 studies assessing chance of death and infarction in patients with coronary collaterals compared to that in patients without.

| | Odds Collaterals | odds no collaterals | n | odds ratio | 95 % ci | z-value | p |
|---|---|---|---|---|---|---|---|
| 1.Monteiro 2003 | 6/29 | 11/24 | 70 | 0.45 | 0.15–1.40 | −1.38 | 1.69 |
| 2.Nathou 2006 | 3/173 | 20/365 | 561 | 0.32 | 0.09–1.08 | −1.84 | 0.066 |
| 3.Meier 2007 | 36/190 | 197/389 | 812 | 0.37 | 0.25–0.56 | −4.87 | 0.0001 |
| 4.Sorajja 2007 | 7/112 | 15/184 | 318 | 0.77 | 0.30–1.94 | −0.56 | 0.576 |
| 5.Regieli 2009 | 7/254 | 16/600 | 879 | 1.03 | 0.42–2.54 | +0.07 | 0.944 |
| 6.Desch 2010 | 5/64 | 34/132 | 235 | 0.30 | 0.11–0.81 | −2.38 | 0.018 |
| 7.Steg 2010 | 246/1676 | 42/209 | 2173 | 0.73 | 0.51–1.04 | −1.72 | 0.085 |

In order to meta-analyze these data, the following calculations are required. OR = odds ratio, lnOR = the natural logarithm of the odds ratio, var = variance.

| | OR | lnOR | var | 1/var | lnOR/var | $(\ln OR)^{2/}$/var |
|---|---|---|---|---|---|---|
| 1.Monteiro 2003 | 0.45 | −0.795 | 0.3337 | 2.997 | −2.382 | 1.894 |
| 2.Nathou 2006 | 0.32 | −1.150 | 0.3919 | 2.882 | −2.935 | 3.375 |
| 3.Meier 2007 | 0.37 | −0.983 | 0.04069 | 24.576 | −24.158 | 23.748 |
| 4.Sorajja 2007 | 0.77 | −0.266 | 0.2239 | 4.466 | −1.188 | 0.3160 |
| 5.Regieli 2009 | 1.03 | −1.194 | 0.2526 | 3.959 | −4.727 | 5.644 |
| 6.Desch 2010 | 0.30 | 0.032 | 0.2110 | 4.739 | 0.152 | 0.005 |
| 7.Stege 2010 | 0.73 | −0.314 | 0.0333 | 30.03 | 9.429 | 2.961 |
| | | | | | | + |
| | | | | 73.319 | −44.667 | 37.943 |

The pooled odds ratio is calculated from antiln of (−44.667/73.319)
= 0.54

(see Chap. 43 for the antiln (anti-logaritm) calculation).

The chi-square value for pooled data $= (-44.667)^2 / 73.319 = 27.2117$

According to the underneath chi-square table for a chi-square value >10.827 and 1 degree of freedom, the p-value

= <0.001

The underneath chi-square table has an upper row with areas under the curve, a left-end column with degrees of freedom, and a whole lot of chi-square values.

Chi-squared distribution

| | Two-tailed *P*-value | | | |
|---|---|---|---|---|
| *df* | 0.10 | 0.05 | 0.01 | 0.001 |
| 1 | 2.706 | 3.841 | 6.635 | 10.827 |
| 2 | 4.605 | 5.991 | 9.210 | 13.815 |
| 3 | 6.251 | 7.851 | 11.345 | 16.266 |
| 4 | 7.779 | 9.488 | 13.277 | 18.466 |
| 5 | 9.236 | 11.070 | 15.086 | 20.515 |
| 6 | 10.645 | 12.592 | 16.812 | 22.457 |
| 7 | 12.017 | 14.067 | 18.475 | 24.321 |
| 8 | 13.362 | 15.507 | 20.090 | 26.124 |
| 9 | 14.684 | 16.919 | 21.666 | 27.877 |
| 10 | 15.987 | 18.307 | 23.209 | 29.588 |
| 11 | 17.275 | 19.675 | 24.725 | 31.264 |
| 12 | 18.549 | 21.026 | 26.217 | 32.909 |
| 13 | 19.812 | 22.362 | 27.688 | 34.527 |
| 14 | 21.064 | 23.685 | 29.141 | 36.124 |
| 15 | 22.307 | 24.996 | 30.578 | 37.698 |
| 16 | 23.542 | 26.296 | 32.000 | 39.252 |
| 17 | 24.769 | 27.587 | 33.409 | 40.791 |
| 18 | 25.989 | 28.869 | 34.805 | 42.312 |
| 19 | 27.204 | 30.144 | 36.191 | 43.819 |
| 20 | 28.412 | 31.410 | 37.566 | 45.314 |
| 21 | 29.615 | 32.671 | 38.932 | 46.796 |
| 22 | 30.813 | 33.924 | 40.289 | 48.268 |
| 23 | 32.007 | 35.172 | 41.638 | 49.728 |
| 24 | 33.196 | 36.415 | 42.980 | 51.179 |
| 25 | 34.382 | 37.652 | 44.314 | 52.619 |
| 26 | 35.536 | 38.885 | 45.642 | 54.051 |
| 27 | 36.741 | 40.113 | 46.963 | 55.475 |

(continued)

| | Two-tailed *P*-value | | | |
|---|---|---|---|---|
| *df* | 0.10 | 0.05 | 0.01 | 0.001 |
| 28 | 37.916 | 41.337 | 48.278 | 56.892 |
| 29 | 39.087 | 42.557 | 49.588 | 58.301 |
| 30 | 40.256 | 43.773 | 50.892 | 59.702 |
| 40 | 51.805 | 55.758 | 63.691 | 73.403 |
| 50 | 63.167 | 67.505 | 76.154 | 86.660 |
| 60 | 74.397 | 79.082 | 88.379 | 99.608 |
| 70 | 85.527 | 90.531 | 100.43 | 112.32 |
| 80 | 96.578 | 101.88 | 112.33 | 124.84 |
| 90 | 107.57 | 113.15 | 124.12 | 137.21 |
| 100 | 118.50 | 124.34 | 135.81 | 149.45 |

## 5 Data Example, Assessing Heterogeneity

The above data will now be assessed for heterogeneity. Heterogeneity of this meta-analysis is tested by the fixed effect model.

Heterogeneity chi-square value = 37.943-27.2117
=10.7317

With 6 degrees of freedom a chi-square value >10.645 the p – value
$= 0.05 < p < 0.10$

Although the meta-analysis shows a significantly lower risk in patients with collaterals than in those without, this result has a limited meaning, since there is a trend to heterogeneity in these studies. For heterogeneity testing it is tested whether the differences between the results of the separate trials are greater than compatible with the play of chance. Additional tests for heterogeneity testing are available (Cleophas and Zwinderman, Meta-analysis. In: Statistics Applied to Clinical Studies, Springer New York, 2012, 5th edition, pp 365–388). When there is heterogeneity in a meta-analysis, a careful investigation of its potential cause is often more important than a lot of additional statistical tests.

## 6 Conclusion

Meta-analyses are systematic reviews of multiple published studies with pooled data. Because the separate studies have different sample sizes a weighted average has to be calculated. Heterogeneity in a meta-analysis means that the differences in the results between the studies are larger than could happen by chance. With a significant heterogeneity the meaning of the pooled data is generally little.

Additional tests for heterogeneity testing are available in Statistics Applied to Clinical Studies 5th edition, Chaps 32–34, Springer New York, 2012). With heterogeneity in a meta-analysis, a careful investigation of its potential cause is important.

# 7 Note

More background, theoretical and mathematical information of meta-analysis given in Statistics applied to clinical studies 5th edition, Chaps. 32–34, Springer Heidelberg Germany, 2012, from the same authors.

# Chapter 58
# Physicians' Daily Life and the Scientific Method

## 1 General Purpose

We assumed the numbers of unanswered questions in the physicians' daily life would be large. But just to get of impression, one of the authors of this work (TC) recorded all of the unanswered answers he asked himself during a single busy day. Excluding the questions with uncertain but generally accepted answers, he included 9 questions.

During the hospital rounds 8.00–12.00 h.

1. Do I continue, stop or change antibiotics with fever relapse after 7 days treatment?
2. Do I prescribe a secondary prevention of a venous thrombosis for 3, 6 months or permanently?
3. Should I stop anticoagulant treatment or continue with a hemorrhagic complication in a patient with an acute lung embolia?
4. Is the rise in falling out of bed lately real or due to chance?
5. Do I perform a liver biopsy or wait and see with liver function disturbance without obvious cause?

During the outpatient clinic 13.00–17.00 h.

6. Do I prescribe aspirin, hydroxy-carbamide or wait and see in a patient with a thrombocytosis of $800 \times 10^{12}$ /l over 6 months?
7. Are fundic gland polyps much more common in females than in males?

During the staff meeting 17.00–18.00 h

8. Is the large number of physicians with burn out due to chance or the result of a local problem?
9. Is the rise in patients' letters of complaints a chance effect or a real effect to worry about?



T.J. Cleophas, A.H. Zwinderman, *Clinical Data Analysis on a Pocket Calculator*,
DOI 10.1007/978-3-319-27104-0_58

Many of the above questions did not qualify for a simple statistical assessment, but others did. The actual assessments, that were very clarifying for our purposes, are given underneath.

## 2 Schematic Overview of Type of Data File

| Outcome binary | Predictor binary |
| --- | --- |
| . | . |
| . | . |
| . | . |
| . | . |
| . | . |
| . | . |
| . | . |
| . | . |
| . | . |
| | |
| | |
| | |
| | |
| | |
| | |

## 3 Primary Scientific Question

Does event analysis in different hospital departments and other patient subgroups provide valuable information for making predictions about health risks and other patient risks.

## 4 Example 1, Falling Out of Bed

If more patients fall out of bed than expected, a hospital department will put much energy in finding the cause and providing better prevention. If, however, the scores tend to rise, another approach is to first assess whether or not the rise is due to chance, because daily life is full of variations. To do so the numbers of events observed is compared the numbers of event in a sister department. The pocket calculator method is a straightforward method for that purpose.

| | Patients with fall out of bed | patients without | |
|---|---|---|---|
| department 1 | 16 (a) | 26 (b) | 42 (a+b) |
| department 2 | 5 (c) | 30 (d) | 35 (c+d) |
| | 21 (a+c) | 56 (b+d) | 77 (a+b+c+d) |

Pocket calculator method:

$$\text{chi-square} = \frac{(ad - bc)^2(a + b + c + d)}{(a + b)(c + d)(b + d)(a + c)} = 5.456.$$

If the chi-square value (see also Chap. 38, and the underneath chi-square table) is larger than 3.841, then a statistically significant difference between the two departments will be accepted at $p < 0.05$. This would mean that in this example, indeed, the difference is larger than could be expected by chance and that a further examination of the measures to prevent fall out of bed is warranted.

The underneath chi-square table has an upper row with areas under the curve, a left-end column with degrees of freedom, and a whole lot of chi-square values.

Chi-squared distribution

| | Two-tailed *P*-value | | | |
|---|---|---|---|---|
| *df* | 0.10 | 0.05 | 0.01 | 0.001 |
| 1 | 2.706 | 3.841 | 6.635 | 10.827 |
| 2 | 4.605 | 5.991 | 9.210 | 13.815 |
| 3 | 6.251 | 7.851 | 11.345 | 16.266 |
| 4 | 7.779 | 9.488 | 13.277 | 18.466 |
| 5 | 9.236 | 11.070 | 15.086 | 20.515 |
| 6 | 10.645 | 12.592 | 16.812 | 22.457 |
| 7 | 12.017 | 14.067 | 18.475 | 24.321 |
| 8 | 13.362 | 15.507 | 20.090 | 26.124 |
| 9 | 14.684 | 16.919 | 21.666 | 27.877 |
| 10 | 15.987 | 18.307 | 23.209 | 29.588 |
| 11 | 17.275 | 19.675 | 24.725 | 31.264 |
| 12 | 18.549 | 21.026 | 26.217 | 32.909 |
| 13 | 19.812 | 22.362 | 27.688 | 34.527 |
| 14 | 21.064 | 23.685 | 29.141 | 36.124 |
| 15 | 22.307 | 24.996 | 30.578 | 37.698 |
| 16 | 23.542 | 26.296 | 32.000 | 39.252 |
| 17 | 24.769 | 27.587 | 33.409 | 40.791 |
| 18 | 25.989 | 28.869 | 34.805 | 42.312 |
| 19 | 27.204 | 30.144 | 36.191 | 43.819 |
| 20 | 28.412 | 31.410 | 37.566 | 45.314 |
| 21 | 29.615 | 32.671 | 38.932 | 46.796 |
| 22 | 30.813 | 33.924 | 40.289 | 48.268 |
| 23 | 32.007 | 35.172 | 41.638 | 49.728 |

(continued)

| Chi-squared distribution | | | | |
|---|---|---|---|---|
| | Two-tailed *P*-value | | | |
| *df* | 0.10 | 0.05 | 0.01 | 0.001 |
| 24 | 33.196 | 36.415 | 42.980 | 51.179 |
| 25 | 34.382 | 37.652 | 44.314 | 52.619 |
| 26 | 35.536 | 38.885 | 45.642 | 54.051 |
| 27 | 36.741 | 40.113 | 46.963 | 55.475 |
| 28 | 37.916 | 41.337 | 48.278 | 56.892 |
| 29 | 39.087 | 42.557 | 49.588 | 58.301 |
| 30 | 40.256 | 43.773 | 50.892 | 59.702 |
| 40 | 51.805 | 55.758 | 63.691 | 73.403 |
| 50 | 63.167 | 67.505 | 76.154 | 86.660 |
| 60 | 74.397 | 79.082 | 88.379 | 99.608 |
| 70 | 85.527 | 90.531 | 100.43 | 112.32 |
| 80 | 96.578 | 101.88 | 112.33 | 124.84 |
| 90 | 107.57 | 113.15 | 124.12 | 137.21 |
| 100 | 118.50 | 124.34 | 135.81 | 149.45 |

## 5 Example 2, Evaluation of Fundic Gland Polyps

A physician has the impression that fundic gland polyps are more common in females than it is in males. Instead of reporting this subjective finding, he decides to follow the next two months every patient in his program.

| | patients with fundic gland polyps | patients without | |
|---|---|---|---|
| females | 15 (a) | 20 (b) | 35 (a + b) |
| males | 15 (c) | 5 (d) | 20 (c + d) |
| | 30 (a + c) | 25 (b + d) | 55 (a + b + c + d) |

Pocket calculator method:

$$\text{chi-square} = \frac{(\text{ad} - \text{bc})^2(\text{a} + \text{b} + \text{c} + \text{d})}{(\text{ab})(\text{c} + \text{d})(\text{b} + \text{d})(\text{a} + \text{c})} = 5.304$$

The calculated chi-square value is again larger than 3.841. The difference between males and females is significant at $p < 0.05$. We can be for about 95 % sure that the difference between the genders is real and not due to chance. The physician can report to his colleagues that the difference in genders is to be taken into account in future work-ups.

## 6 Example 3, Physicians with a Burn-Out

Two partnerships of specialists have the intention to associate. However, during meetings, it was communicated that in one of the two partnerships there were three specialists with burn-out. The meeting decided not to consider this as chance finding, but requested a statistical analysis of this finding under the assumption that unknown factors in partnership 1 may place these specialists at an increased risk of a burn-out.

| | physicians with burn out | without burn out | |
|---|---|---|---|
| partnership 1 | 3 (a) | 7 (b) | 10 (a + b) |
| partnership 2 | 0 (c) | 10 (d) | 10 (c + d) |
| | 3 (a + c) | 17(b + d) | 20 (a + b + c + d) |

pocket calculator method

$$\text{chi-square} = \frac{(ad-bc)^2(a+b+c+d)}{(a+b)(c+d)(b+d)(a+c)} = \frac{(30-0)2\ (20)}{10\times10\times17\times3} = \frac{900\times20}{\ldots\ldots\ldots} = 3.6$$

The chi-square value was between 2.706 and 3.841. This means that no significant difference between the two partnerships exists, but there is a trend to a difference at $p < 0.10$. This was communicated back to the meeting and it was decided to disregard the trend. Ten years later no further case of burn-out had been observed.

# 7 Example 4, Patients' Letters of Complaints

In a hospital the number of patients' letters of complaints was twice the number in the period before. The management was deeply worried and issued an in-depth analysis of possible causes. One junior manager recommended that prior to this laborious exercise it might be wise to first test whether the increase might be due to chance rather than a real effect.

| | patients with letter of complaints | patients without | |
|---|---|---|---|
| year 2006 | 10 (a) | 1000 (b) | 1010 (a + b) |
| year 2005 | 5 (c) | 1000 (d) | 1005 (c + d) |
| | 15 (a + c) | 2000 (b + d) | 2015 (a + b + c + d) |

$$\text{chi-square} = \frac{(ad - bc)^2(a + b + c + d)}{(a + b)(c + d)(b + d)(a + c)} = 1.64$$

The chi-square was smaller than 2.706, and so the difference could not be ascribed to any effect to worry about but rather to chance. No further analysis of the differences between 2006 and 2005 were performed.

# 8 Conclusion

There are, of course, many questions in physicians' daily life that are less straightforward and cannot be readily answered at the workplace with a pocket calculator. E.g., the effects of subgroups and other covariates in a patient group will require t-tests, analyses of variance, likelihood ratio tests, and regression models. Fortunately, in the past 15 years user-friendly statistical software and self-assessment programs have been developed that can help answering complex questions. The complementary titles of this book, entitled Statistics applied to clinical studies 5th edition, machine in medicine a complete overview, and SPSS for starters and second levels, Springer Heidelberg Germany, 2012–2015, from the same authors, are helpful for the purpose.

So far, few physicians have followed the scientific method for answering practical questions they, simply, do not know the answer to. *The scientific method can be summarized in a nutshell: reformulate your question into a hypothesis, and try and test this hypothesis against control observations.*

# 9 Note

More background, theoretical and mathematical information of the scientific method during daily life is given in Statistics applied to clinical studies 5th edition, Chap. 60, Springer Heidelberg Germany, 2012, from the same authors.

# Chapter 59
# Incident Analysis and the Scientific Method

## 1 General Purpose

The PRISMA (Prevention and Recovery System for Monitoring and Analysis) –, CIA (Critical Incident Analysis) –, CIT (Critical Incident Technique) –, TRIPOD (tripod-theory based method) – methods are modern approaches to incident – analysis. It is unclear why the scientific method has been systematically ignored in incident – analysis. As example the case of a fatal hemorrhage in a hospital during an observational period of one year was used. In case of a fatal hemorrhage the physician in charge of the analysis will first make an inventory of how many fatal hemorrhages of the same kind have occurred in the period of one year. The number seems to be no less than ten. The question is, is this number larger than could happen by chance. This chapter assesses the use of z-tests and chi-square tests for answering the question of chance findings.

## 2 Schematic Overview of Type of Data File

| Outcome binary |
|---|
| . |
| . |
| . |
| . |
| . |
| . |
| . |
| . |
| |
| |
| |



T.J. Cleophas, A.H. Zwinderman, *Clinical Data Analysis on a Pocket Calculator*,
DOI 10.1007/978-3-319-27104-0_59

## 3 Primary Scientific Question

In clinical facilities medical events are an important criterion of quality of care, and careful analysis is non plus ultra. Can simple statistical tests be helpful for that purpose.

## 4 Data Example, Test 1

As example the case of a fatal hemorrhage in a hospital during an observational period of one year was used. In case of a fatal hemorrhage the physician in charge of the analysis will first make an inventory of how many fatal hemorrhages of the same kind have occurred in the period of one year. The number seems to be ten.

The null - hypothesis is that 0 hemorrhages will occur per year, and the question is whether 10 is significantly more than 0. A one – sample – z - test is used.

$z = (\text{mean number})/(\text{standard error}) = 10/\sqrt{10} = 3.16$.
z – value is larger than 3.080 (see underneath t-table).
p – value is $< 0.002$.

The number 10 is, thus, much larger than a number that could occur by accident. Here an avoidable error could very well be responsible. However, a null - hypothesis of 0 hemorrhages is probably not correct, because a year without fatal hemorrhages, actually, never happens. Therefore, we will compare the number of fatal hemorrhages in the given year with that of the year before. There were five fatal hemorrhages then. The z - test produces the following result (see underneath t-table).

T-Table

| df | One-Tail = .4<br>Two-Tail = .8 | .25<br>.5 | .1<br>.2 | .05<br>.1 | .025<br>.05 | .01<br>.02 | .005<br>.01 | .0025<br>.005 | .001<br>.002 | .0005<br>.001 |
|---|---|---|---|---|---|---|---|---|---|---|
| 1 | 0.325 | 1.000 | 3.078 | 6.314 | 12.706 | 31.821 | 63.657 | 127.32 | 318.31 | 636.62 |
| 2 | 0.289 | 0.816 | 1.886 | 2.920 | 4.303 | 6.965 | 9.925 | 14.089 | 22.327 | 31.598 |
| 3 | 0.277 | 0.765 | 1.638 | 2.353 | 3.182 | 4.541 | 5.841 | 7.453 | 10.214 | 12.924 |
| 4 | 0.271 | 0.741 | 1.533 | 2.132 | 2.776 | 3.747 | 4.604 | 5.598 | 7.173 | 8.610 |
| 5 | 0.267 | 0.727 | 1.476 | 2.015 | 2.571 | 3.365 | 4.032 | 4.773 | 5.893 | 6.869 |
| 6 | 0.265 | 0.718 | 1.440 | 1.943 | 2.447 | 3.143 | 3.707 | 4.317 | 5.208 | 5.959 |
| 7 | 0.263 | 0.711 | 1.415 | 1.895 | 2.365 | 2.998 | 3.499 | 4.029 | 4.785 | 5.408 |
| 8 | 0.262 | 0.706 | 1.397 | 1.860 | 2.306 | 2.896 | 3.355 | 3.833 | 4.501 | 5.041 |
| 9 | 0.261 | 0.703 | 1.383 | 1.833 | 2.262 | 2.821 | 3.250 | 3.690 | 4.297 | 4.781 |
| 10 | 0.260 | 0.700 | 1.372 | 1.812 | 2.228 | 2.764 | 3.169 | 3.581 | 4.144 | 4.587 |
| 11 | 0.260 | 0.697 | 1.363 | 1.796 | 2.201 | 2.718 | 3.106 | 3.497 | 4.025 | 4.437 |
| 12 | 0.259 | 0.695 | 1.356 | 1.782 | 2.179 | 2.681 | 3.055 | 3.428 | 3.930 | 4.318 |
| 13 | 0.259 | 0.694 | 1.350 | 1.771 | 2.160 | 2.650 | 3.012 | 3.372 | 3.852 | 4.221 |
| 14 | 0.258 | 0.692 | 1.345 | 1.761 | 2.145 | 2.624 | 2.977 | 3.326 | 3.787 | 4.140 |
| 15 | 0.258 | 0.691 | 1.341 | 1.753 | 2.131 | 2.602 | 2.947 | 3.286 | 3.733 | 4.073 |
| 16 | 0.258 | 0.690 | 1.337 | 1.746 | 2.120 | 2.583 | 2.921 | 3.252 | 3.686 | 4.015 |
| 17 | 0.257 | 0.689 | 1.333 | 1.740 | 2.110 | 2.567 | 2.898 | 3.222 | 3.646 | 3.965 |
| 18 | 0.257 | 0.688 | 1.330 | 1.734 | 2.101 | 2.552 | 2.878 | 3.197 | 3.610 | 3.922 |
| 19 | 0.257 | 0.688 | 1.328 | 1.729 | 2.093 | 2.539 | 2.861 | 3.174 | 3.579 | 3.883 |
| 20 | 0.257 | 0.687 | 1.325 | 1.725 | 2.086 | 2.528 | 2.845 | 3.153 | 3.552 | 3.850 |
| 21 | 0.257 | 0.686 | 1.323 | 1.721 | 2.080 | 2.518 | 2.831 | 3.135 | 3.527 | 3.819 |
| 22 | 0.256 | 0.686 | 1.321 | 1.717 | 2.074 | 2.508 | 2.819 | 3.119 | 3.505 | 3.792 |
| 23 | 0.256 | 0.685 | 1.319 | 1.714 | 2.069 | 2.500 | 2.807 | 3.104 | 3.485 | 3.767 |
| 24 | 0.256 | 0.685 | 1.318 | 1.711 | 2.064 | 2.492 | 2.797 | 3.091 | 3.467 | 3.745 |
| 25 | 0.256 | 0.684 | 1.316 | 1.708 | 2.060 | 2.485 | 2.787 | 3.078 | 3.450 | 3.725 |
| 26 | 0.256 | 0.684 | 1.315 | 1.706 | 2.056 | 2.479 | 2.779 | 3.067 | 3.435 | 3.707 |
| 27 | 0.256 | 0.684 | 1.314 | 1.703 | 2.052 | 2.473 | 2.771 | 3.057 | 3.421 | 3.690 |
| 28 | 0.256 | 0.683 | 1.313 | 1.701 | 2.048 | 2.467 | 2.763 | 3.047 | 3.408 | 3.674 |
| 29 | 0.256 | 0.683 | 1.311 | 1.699 | 2.045 | 2.462 | 2.756 | 3.038 | 3.396 | 3.659 |
| 30 | 0.256 | 0.683 | 1.310 | 1.697 | 2.042 | 2.457 | 2.750 | 3.030 | 3.385 | 3.646 |
| 40 | 0.255 | 0.681 | 1.303 | 1.684 | 2.021 | 2.423 | 2.704 | 2.971 | 3.307 | 3.551 |
| 60 | 0.254 | 0.679 | 1.296 | 1.671 | 2.000 | 2.390 | 2.660 | 2.915 | 3.232 | 3.460 |
| 120 | 0.254 | 0.677 | 1.289 | 1.658 | 1.980 | 2.358 | 2.617 | 2.860 | 3.160 | 3.373 |
| ∞ | 0.253 | 0.674 | 1.282 | 1.645 | 1.960 | 2.326 | 2.576 | 2.807 | 3.090 | 3.291 |

The left-end column of the above t-table gives degrees of freedom (≈ sample sizes), two top rows with p-values (areas under the curve), and the t-table is, furthermore, full of t-values, that, with ∞ degrees of freedom, become equal to z-values.

$z = (10-5)/\surd(10+5) = 1.29$

p – value = not significant, because z is $<1.96$.

We can, however, question whether both years are representative for a longer period of time. Epidemiological data have established that an incident-reduction of 70% is possible with optimal quality health care. We test whether 10 is significantly different from (10 - (70%) x 10) = 3. The z-test shows the following.

$z = (10-3)/\sqrt{(10+3)} = 1.94$
p-value $= 0.05 < p < 0.10$

It means, that here also no significant effect has been demonstrated. A more sensitive mode of testing will be obtained, if we take into account the entire number of admissions per year. In the given hospital there were 10,000 admissions in either of the two years. A chi-square test can now be performed.

## 5 Data Example, Test 2

A chi-square test can now be performed according to the 2 x 2 contingency table in the underneath table.

With one degree of freedom this value ought to have been at least 3.84 in order to demonstrate whether a significant difference is in the data (chi-square table). And, so, again there is no significant difference between the two years.

Rates of fatal hemorrhages in a hospital during two subsequent year of observation. According to the chi - square statistic < 3.84 the difference in rates is not significant

| | Year 1 | Year 2 |
|---|---|---|
| Number fatal hemorrhages | 10 | 5 |
| Number control patients | 9990 | 9995 |

$$\text{Chi - square} = \frac{(10 \times 9995 + 5 \times 9990)^2\ (20000)}{10 \times 9990 \times 5 \times 9995} = 1.62$$

The underneath chi-square table has an upper row with areas under the curve, a left-end column with degrees of freedom, and a whole lot of chi-square values.

Chi-squared distribution

| | Two-tailed *P*-value | | | |
|---|---|---|---|---|
| *df* | 0.10 | 0.05 | 0.01 | 0.001 |
| 1 | 2.706 | 3.841 | 6.635 | 10.827 |
| 2 | 4.605 | 5.991 | 9.210 | 13.815 |
| 3 | 6.251 | 7.851 | 11.345 | 16.266 |
| 4 | 7.779 | 9.488 | 13.277 | 18.466 |
| 5 | 9.236 | 11.070 | 15.086 | 20.515 |
| 6 | 10.645 | 12.592 | 16.812 | 22.457 |
| 7 | 12.017 | 14.067 | 18.475 | 24.321 |
| 8 | 13.362 | 15.507 | 20.090 | 26.124 |

(continued)

| *df* | Two-tailed *P*-value 0.10 | 0.05 | 0.01 | 0.001 |
|---|---|---|---|---|
| 9 | 14.684 | 16.919 | 21.666 | 27.877 |
| 10 | 15.987 | 18.307 | 23.209 | 29.588 |
| 11 | 17.275 | 19.675 | 24.725 | 31.264 |
| 12 | 18.549 | 21.026 | 26.217 | 32.909 |
| 13 | 19.812 | 22.362 | 27.688 | 34.527 |
| 14 | 21.064 | 23.685 | 29.141 | 36.124 |
| 15 | 22.307 | 24.996 | 30.578 | 37.698 |
| 16 | 23.542 | 26.296 | 32.000 | 39.252 |
| 17 | 24.769 | 27.587 | 33.409 | 40.791 |
| 18 | 25.989 | 28.869 | 34.805 | 42.312 |
| 19 | 27.204 | 30.144 | 36.191 | 43.819 |
| 20 | 28.412 | 31.410 | 37.566 | 45.314 |
| 21 | 29.615 | 32.671 | 38.932 | 46.796 |
| 22 | 30.813 | 33.924 | 40.289 | 48.268 |
| 23 | 32.007 | 35.172 | 41.638 | 49.728 |
| 24 | 33.196 | 36.415 | 42.980 | 51.179 |
| 25 | 34.382 | 37.652 | 44.314 | 52.619 |
| 26 | 35.536 | 38.885 | 45.642 | 54.051 |
| 27 | 36.741 | 40.113 | 46.963 | 55.475 |
| 28 | 37.916 | 41.337 | 48.278 | 56.892 |
| 29 | 39.087 | 42.557 | 49.588 | 58.301 |
| 30 | 40.256 | 43.773 | 50.892 | 59.702 |
| 40 | 51.805 | 55.758 | 63.691 | 73.403 |
| 50 | 63.167 | 67.505 | 76.154 | 86.660 |
| 60 | 74.397 | 79.082 | 88.379 | 99.608 |
| 70 | 85.527 | 90.531 | 100.43 | 112.32 |
| 80 | 96.578 | 101.88 | 112.33 | 124.84 |
| 90 | 107.57 | 113.15 | 124.12 | 137.21 |
| 100 | 118.50 | 124.34 | 135.81 | 149.45 |

## 6 Data Example, Test 3

Finally, a log – likelihood – ratio - test will be performed, a test which falls into the category of exact - tests, and is, generally, still somewhat more sensitive (see also Chap. 46). The result is in the table below. It is close to 3.84, but still somewhat smaller.

Log likelihood ratio test of the data is below. Also this test is not significant with a chi-square value smaller than 3.84

$$\text{Log likelihood ratio} = 5 \log \frac{(10/9990)}{5/9995} + 9995 \log \frac{(1 - 10/9990)}{1 - 5/9995}$$

$$= 3.468200 - 5.008856 = -1.540656$$

$$\text{Chi - square} = -2 \log - \text{likelihood} - \text{ratio} = 3.0813$$

log = natural logarithm

Also the above test shows no significant difference between the frequencies of deadly fatal hemorrhages in the two years of observation.

## 7 Data Example, Test 4

The analyst in charge takes the decision to perform one last test, making use of epidemiological data that have shown that with optimal health care quality in a facility similar to ours we may accept with 95% confidence that the number of fatal hemorrhages will remain below 20 per 10,000 admissions. With 10 deadly bleedings the 95% confidence interval can be calculated to be 5–18 (calculated from the Internet "Confidence interval calculator for proportions", http://faculty.vassar.edu). This result is under 20. Also from this analysis it can be concluded that a profound research of the fatal hemorrhages is not warranted. The number of hemorrhages falls under the boundary of optimal quality care.

## 8 Conclusion

The scientific method is often defined as an evaluation of clinical data based on appropriate statistical tests, rather than a description of the cases and their summaries. The above example explains that the scientific method can be helpful in giving a clue to which incidents are based on randomness and which are not so.

Many incident analysis software programs provide systematic approaches to the explanation of a single incident at the workplace. Few such software programs have followed the scientific method for assessing causal relationships between factors and the resulting incident. *The scientific method is in a nutshell: reformulate your question into a hypothesis and try and test this hypothesis against control observations.*

# 9 Note

More background, theoretical and mathematical information of incident analysis and the scientific method is given in Statistics applied to clinical studies 5th edition, Chap. 61, Springer Heidelberg Germany, 2012, from the same authors.

# Chapter 60
# Cochran Q-Test for Large Paired Cross-Tabs

## 1 General Purpose

To analyze samples of more than 2 pairs of data, e.g., 3, 4 pairs, etc., McNemar's test (Chap. 48 and 49) can not be applied. For that purpose Cochran's test or logistic regression analysis is adequate (Chap. 45). The current chapter assesses how the Cochran's test works.

## 2 Schematic Overview of Type of Data File

| Outcome (binary) predictor 1 | Outcome predictor 2 | Outcome predictor 3 |
|---|---|---|
| . | . | |
| . | . | . |
| . | . | . |
| . | . | . |
| . | . | . |
| . | . | . |
| . | . | . |
| . | . | . |
| . | . | . |
| | | |
| | | |
| | | |
| | | |
| | | |
| | | |



T.J. Cleophas, A.H. Zwinderman, *Clinical Data Analysis on a Pocket Calculator*,
DOI 10.1007/978-3-319-27104-0_60

## 3 Primary Scientific Question

Is the Cochran Q-test able to tell the differences between three or more paired observations of yes / no responders to three or more different treatments.

## 4 Cochran Q-Test for a Study with Three Paired Binary Observations per Patient

The underneath table shows three paired observations in each single patient. The paired property of these observations has to be taken into account because of the, generally, positive correlation between paired observations. Cochran's Q test is appropriate for that purpose.

Responders and non-responders to three different treatments.

(treat = treatment, 1 = responder, 0 = non-responder)

| Patient | treat1 | treat2 | treat3 |
|---|---|---|---|
| 1 | 1 | 0 | 0 |
| 2 | 0 | 0 | 1 |
| 3 | 0 | 0 | 1 |
| 4 | 0 | 0 | 1 |
| 5 | 0 | 0 | 1 |
| 6 | 1 | 0 | 0 |
| 7 | 1 | 0 | 0 |
| 8 | 1 | 0 | 0 |
| 9 | 1 | 0 | 0 |
| 10 | 0 | 0 | 1 |
| 11 | 0 | 0 | 1 |
| 12 | 0 | 0 | 1+ |
| | 5 | 0 | 7 |

Grans mean columns

$$= (5 + 0 + 7)/3 = 12/3 = 4$$

$$\text{Chi-square} = 3(3-1) \times \left(\sum \text{columns}^2 / \sum \text{rows}\right)$$

$$\sum \text{columns}^2 = (5-4)^2 + (0-4)^2 + (7-4)^2 = 1 + 16 + 9$$

$$= 26$$

∑ rows =
for row 1 (number treatments) – (number responders) = 3 – 1 = 2
for row 2 = 2
for row 3 = 2
for row 4 = 2
for row 5 = 2
for row 6 = 2
for row 7 = 2
for row 8 = 2
for row 9 = 2
for row 10 = 2
for row 11 = 2
for row 12 = 2+
= 24

Chi-square = 6 × 26/24 = 6.5

With 3 treatments we have (3 – 1) degrees of freedom.

The underneath chi-square table has an upper row with areas under the curve, a left-end column with degrees of freedom, and a whole lot of chi-square values. It show that for 2 degrees of freedom, a chi-square value >5.991 means that $p < 0.05$. A significant difference between the patterns of responding to the treatments 1, 2, and 3.

Chi-squared distribution

| | Two-tailed *P*-value | | | |
|---|---|---|---|---|
| *df* | 0.10 | 0.05 | 0.01 | 0.001 |
| 1 | 2.706 | 3.841 | 6.635 | 10.827 |
| 2 | 4.605 | 5.991 | 9.210 | 13.815 |
| 3 | 6.251 | 7.851 | 11.345 | 16.266 |
| 4 | 7.779 | 9.488 | 13.277 | 18.466 |
| 5 | 9.236 | 11.070 | 15.086 | 20.515 |
| 6 | 10.645 | 12.592 | 16.812 | 22.457 |
| 7 | 12.017 | 14.067 | 18.475 | 24.321 |
| 8 | 13.362 | 15.507 | 20.090 | 26.124 |
| 9 | 14.684 | 16.919 | 21.666 | 27.877 |
| 10 | 15.987 | 18.307 | 23.209 | 29.588 |
| | | | | |
| 11 | 17.275 | 19.675 | 24.725 | 31.264 |
| 12 | 18.549 | 21.026 | 26.217 | 32.909 |
| 13 | 19.812 | 22.362 | 27.688 | 34.527 |
| 14 | 21.064 | 23.685 | 29.141 | 36.124 |
| 15 | 22.307 | 24.996 | 30.578 | 37.698 |
| 16 | 23.542 | 26.296 | 32.000 | 39.252 |
| 17 | 24.769 | 27.587 | 33.409 | 40.791 |
| 18 | 25.989 | 28.869 | 34.805 | 42.312 |

(continued)

| | Two-tailed $P$-value | | | |
|---|---|---|---|---|
| *df* | 0.10 | 0.05 | 0.01 | 0.001 |
| 19 | 27.204 | 30.144 | 36.191 | 43.819 |
| 20 | 28.412 | 31.410 | 37.566 | 45.314 |
| | | | | |
| 21 | 29.615 | 32.671 | 38.932 | 46.796 |
| 22 | 30.813 | 33.924 | 40.289 | 48.268 |
| 23 | 32.007 | 35.172 | 41.638 | 49.728 |
| 24 | 33.196 | 36.415 | 42.980 | 51.179 |
| 25 | 34.382 | 37.652 | 44.314 | 52.619 |
| 26 | 35.536 | 38.885 | 45.642 | 54.051 |
| 27 | 36.741 | 40.113 | 46.963 | 55.475 |
| 28 | 37.916 | 41.337 | 48.278 | 56.892 |
| 29 | 39.087 | 42.557 | 49.588 | 58.301 |
| 30 | 40.256 | 43.773 | 50.892 | 59.702 |
| 40 | 51.805 | 55.758 | 63.691 | 73.403 |
| 50 | 63.167 | 67.505 | 76.154 | 86.660 |
| 60 | 74.397 | 79.082 | 88.379 | 99.608 |
| 70 | 85.527 | 90.531 | 100.43 | 112.32 |
| 80 | 96.578 | 101.88 | 112.33 | 124.84 |
| 90 | 107.57 | 113.15 | 124.12 | 137.21 |
| 100 | 118.50 | 124.34 | 135.81 | 149.45 |

Mc Nemar tests (Chaps. 48 and 49) must be performed to find out, where the differences are: between treatments 1 and 2, 2 and 3, and/or 1 and 3. Adjustment for multiple testing is also required (Chap.18).

## 5 Conclusion

To analyze samples of more than 2 pairs of yes-no data, e.g., 3, 4 paired columns, McNemar's test can not be applied. For that purpose Cochran's test or logistic regression analysis is adequate. The current chapter assesses how the Cochran test works. Like with paired analysis of variance (Chap. 20), Post hoc analyses, and adjustment for multiple testing are required.

## 6 Note

More background, theoretical and mathematical information of Cochran Q-test is given in Statistics applied to clinical studies 5th edition, Chap. 3, Springer Heidelberg Germany, 2012, from the same authors.

# Index




T.J. Cleophas, A.H. Zwinderman, *Clinical Data Analysis on a Pocket Calculator*,
DOI 10.1007/978-3-319-27104-0

Printed in the United States
By Bookmasters